The Electronic Packaging Series

Series Editor: Michael G. Pecht, University of Maryland

Advanced Routing of Electronic Modules
Michael Pecht and Yeun Tsun Wong

Electronic Packaging Materials and Their Properties
Michael Pecht, Rakesh Agarwal, Patrick McCluskey, Terrance Dishongh, Sirus Javadpour, and Rahul Mahajan

Guidebook for Managing Silicon Chip Reliability
Michael Pecht, Riko Radojcic, and Gopal Rao

High Temperature Electronics
Patrick McCluskey, Thomas Podlesak, and Richard Grzybowski

Influence of Temperature on Microelectronics and System Reliability
Pradeep Lall, Michael Pecht, and Edward Hakim

Long-Term Non-Operating Reliability of Electronic Products
Michael Pecht and Judy Pecht

Guidebook for Managing Silicon Chip Reliability

Michael G. Pecht
Riko Radojcic
Gopal Rao

CRC Press
Boca Raton London New York Washington, D.C.

Library of Congress Cataloging-in-Publication Data

Pect, Michael.

Guidebook for managing silicon chip reliability / Michael G. Pecht, Riki Radojcic, Gopal Rao.

p. cm. -- (Electronic packaging series)

Includes bibliographical references and index.

ISBN 0-8493-9624-7 (alk. paper)

1. Integrated circuits--Reliability. 2. Semiconductors--Reliability. 3. Silicon. 4. Electronic packaging--Reliability.sili. I. Radojcic, Riko. II. Rao, Gopal K. III. Title. IV. Series.

TK7874.P425 1998

621.3815—dc21 98-24467

CIP

International Standard Book Number 0-8493-9624-7

Library of Congress Card Number 98-24467

Printed in the United States of America 1 2 3 4 5 6 7 8 9 0

Printed on acid-free paper

Preface

Modern systems, containing many millions of transistors, are, by and large, very reliable, and tend to perform far better than was thought possible only a few years ago. The key question is what happens next – how is the understanding of the failure mechanisms and experiences gained to date going to be used in the future? It is clear that laws of physics will apply to future products too, and in fact that they will have to be 'more strictly enforced.' It is also clear that some of the practices from the past will have to be changed to deal with the future.

Deep sub-micron technologies will aggravate many of the reliability mechanisms and the way that they are managed. However, the physical foundation behind most of these trends is based on increasing chip complexity, increasing chip operating frequency, and decreasing power supply voltage. These factors are the source of the technical concerns posed by the deep sub-micron technologies.

Achieving cost-effective and timely performance over time requires an organized, disciplined, and time-phased approach to product design, development, qualification, manufacture, and in-service management.

Managing typical specifications that define chip performance, cost, schedule, quality, and reliability becomes significantly more difficult with deep sub-micron chip complexities. It is clear that the drive throughout the industry is towards developing design tools and practices that manage this balance better.

This book focuses on the issues associated with the individual failure mechanisms associated with semiconductor devices. The first chapter is an introduction. Chapter 2 presents a list of failure sites, operational loads, and failure mechanisms associated with semiconductor devices. Chapter 3 focuses on intrinsic device sensitivities. Chapter 4 presents electromigration. Hot-carrier aging is presented in Chapter 5, followed by time-dependent dielectric breakdown in Chapter 6. Details of the mechanical stress induced migration are presented in Chapter 7. Alpha particle sensitivity is presented in Chapter 8. Chapter 9 presents electrostatic discharge (ESD) and electrical overstress and is followed by Chapter 10 that focuses on latchup. Chapter 11 presents some guidelines for designing for reliability. Chapter 12 discusses process development issues and Chapter 13 some manufacturing data. Chapter 14 discusses qualification issues and Chapter 15 provides insight into screening. Finally, Chapter 16 presents a summary of the topics in the book.

The material presented in this book is the result of the efforts of two dedicated teams: one at the University of Maryland CALCE Electronic Packaging Research Center, and one at Cadence CSD (formerly Unisys). The authors merely served the purpose of articulating and distilling the understandings derived by these two teams. It is interesting to note that the two teams arrived at similar conclusions from two different starting points. CALCE's efforts originated in exploring the reliability of electronic systems, and reached down into the physics of failures associated with the systems, subsystems and ICs. Cadence's (Unisys) efforts originated in efforts to manufacture reliable ICs and reached up to exploring the impact of system applications on chip reliability. In spite of, or maybe because of, the differences of the starting points, the authors have found a singular agreement in outlook and philosophy for managing chip reliability.

Besides thanking the teams, there are a few individuals that deserve special note and recognition. The outstanding members of the Cadence team that need to be acknowledged and thanked include: Lance Oshiro, Michael Zaragoza, and Mark Nakamoto. In addition, the enlightened Cadence and Unisys management teams that supported and nurtured this work should also be recognized.

The outstanding members of the CALCE team that need to be acknowledged and thanked include Drs. Donald Barker, Patrick McCluskey, Abhijit Dasgupta, and Yogi Joshi, of CALCE Electronic Products and Systems Consortium at the University of Maryland, and Dr. Asaf Katz of Israel.

Michael Pecht is the Director of the CALCE Electronic Packaging Research Center (EPRC) at the University of Maryland and a Full Professor with a three-way joint appointment in Mechanical Engineering, Engineering Research, and Systems Research. Dr. Pecht has a B.S. in Acoustics, a M.S. in Electrical Engineering, and a M.S. and Ph.D. in Engineering Mechanics from the University of Wisconsin. He is a Professional Engineer, an IEEE Fellow, an ASME Fellow and a Westinghouse Fellow. He has written eleven books on electronics products development. He served as chief editor of the IEEE Transactions on Reliability for eight years and on the advisory board of IEEE Spectrum. He is currently the chief editor for Microelectronics and Reliability International, an associate editor for the IEEE Transactions on Components, Packaging, and Manufacturing Technology; SAE Reliability, Maintainability and Supportability Journal; and the International Microelectronics Journal, and on the advisory board of the Journal of Electronics Manufacturing. He serves on the board of advisors for various companies and consults for the U.S. government, providing expertise in strategic planning in the area of electronics products development and marketing.

Riko Radojcic, currently Director of Silicon Device Technology activity at Cadence Design Systems CSD branch, has been working in the semiconductor industry for 19 years. He received his B.Sc. and Ph.D. degrees in Electrical Engineering from the University of Salford in the U.K. Since then he has been a Process Engineer with Ferranti Electronics, in UK, and has held a number of positions with Unisys Corp. (Burroughs Corp.) in the area of silicon semiconductor devices, manufacturing process characterization and modeling, and IC reliability engineering. He has pioneered and implemented in a real industrial environment various methodologies for managing IC reliability.

Gopal Rao is currently working at Intel Corporation, Albuquerque, New Mexico. He is currently the world-wide Quality and Reliability group manager for all the high volume microprocessor fabrication operations at Intel. Gopal has been in the semiconductor industry for 17 years and has held various engineering and management positions in high technology companies both in the USA and Europe. He has been recognized for his contributions in the field of chip interconnect technology. Specifically, he received Intel's highest award for his effort in new process and product qualification cycle time reduction and for accelerating product time to market. Gopal is the author of a widely accepted book, "Multilevel Interconnect Technology", published by Intel/McGraw Hill. Gopal holds two Master of Science degrees in Physics and Material Science.

CONTENTS

1. INTRODUCTION

During the infancy of the modern semiconductor industry, the 'conventional belief' was that solid state components would be free of any reliability issues, due to the absence of any 'moving parts.' Experience over the last few decades has shown this belief to be overly optimistic. Management of reliability is one of the key semiconductor engineering concerns, especially when dealing with complex high component counts, or systems subject to extreme environmental conditions. Consequently, over the years, the industry has evolved a set of disciplines, and a specialty field of study to address reliability. This book describes the principal failure mechanisms associated with modern ICs, and the typical practices used to address them.

The 'conventional wisdom' evolved over the last few decades dictates that the reliability characteristics of IC components can be described by the 'bathtub' curve. Infant mortalities (decreasing failure rate at the front end of the curve) are caused by manufacturing defects and/or specific design sensitivities. Random failures (constant failure rate at the bottom of the bath tub) are acts of God (e.g., EOS from lightening), or man (e.g., EOS/ESD due to handling), or are caused by interactions with process defects. Wearout (increasing failure rate at the back end of the curve) is caused by intrinsic material wear and/or fatigue.

The 'conventional methodology' for management of IC reliability dictates that component reliability is to be empirically demonstrated through accelerated life tests – i.e., tests at elevated temperature or voltage, extreme mechanical or thermal cycling. In fact, a set of 'standard' tests has evolved and defined in military specifications and standards. If a reliability issue is found during these tests, the circuit is redesigned or the manufacturing process is changed, and the product is tested again. However, this sequential approach increasingly does not satisfy modern IC schedule, cost, quality and reliability requirements. Furthermore, technology trends indicating continued increases in chip complexity, decreases in process defectivity and chip counts per system, all dictate a need for moving the reliability management practices beyond the current 'standard' approaches.

Best-in-class companies are consequently developing a set of new practices, based on the physics-of-failures, reliability statistics, and an understanding of interactions between the design and process attributes, in order to manage reliability throughout the IC product life cycle.

2. HOW DEVICES FAIL

A failure mechanism is defined as the chemical, electrical, physical, mechanical, or thermal process leading to failure. Failure mechanisms in microelectronics devices are either categorized according to the type, or mode, of the failure precipitated, or according to the root cause, or mechanism. Failure modes include 'shorts', 'opens,' or parametric drifts. Failure mechanisms address the physical phenomena such as overstress, electromigration, and corrosion.

Failure mechanisms can also be categorized by the nature of the loads—mechanical, thermal, electrical, radiation, and chemical—that trigger or accelerate the mechanism. For example, mechanical failures can result from elastic or plastic deformation, buckling, brittle or ductile fracture, interfacial separation, fatigue crack initiation and propagation, creep, and creep rupture. Thermal failures can arise when the component is operated beyond the critical temperature of some component material (such as the glass-transition temperature, the melting point, or flash point), or by severely denaturing the material interfaces creating intermetallics or diffused interfaces. Electrical failures include those due to electrostatic discharge, dielectric breakdown, junction breakdown, hot electron injection, surface and bulk trapping, surface breakdown, and electromigration. Radiation failures are caused principally by uranium and thorium contaminants associated with some of the materials used or by secondary cosmic rays. Chemical failures arise in environments that accelerate corrosion, oxidation, and ionic surface dendritic growth.

Different types of loads can also interact to cause failure. For example, a thermal load can trigger mechanical failure because of a thermal expansion mismatch. Other interactive failures include stress-assisted corrosion, stress-corrosion cracking, field-induced metal migration, and temperature-induced acceleration of chemical reactions. It has been predicted that the supply voltage used in state-of-art circuits will be 0.9V by the year 2004. For deep submicron devices, low supply voltage should lead to higher performance. Nevertheless, an aggressive low voltage/low power technology will alter many traditional reliability problems such as metallization failure, oxide breakdown, hot carrier effects, electrostatic discharge, leakage currents, circuit noise and alpha particle sensitivity. One of the consequences of continued scaling of dimensions is to increase this sensitivity to defects; hence defect control in modern sub-micron processes is that much more important. The design team must be aware of all potential failure mechanisms at the various device sites (Figure 2.1).

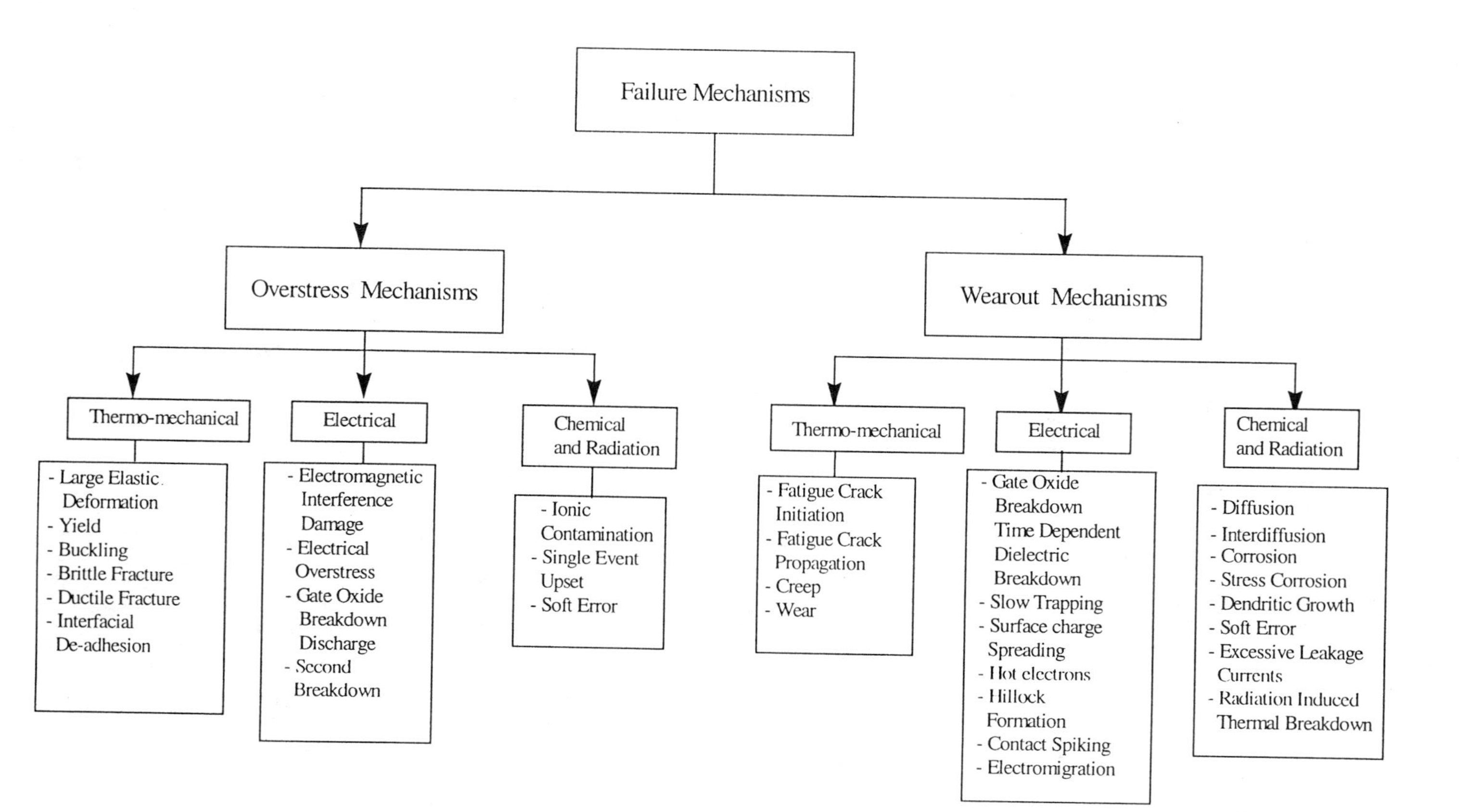

Figure 2-1 Classification of failure mechanisms. Interactions are also possible and combinations of conditions can lead to accelerated failure.

2.1 Intrinsic Mechanisms

'Intrinsic' failures are defined as inherent in the design and materials used, rather than being caused by an interaction of a load or stress with some defect. Intrinsic failures are typically wearout type failure mechanisms. Intrinsic failures are typically managed rather than eliminated – by insuring that wearout occurs beyond the useful life of a product. Generally, this is accomplished through a set of 'design rules' that dictate limits on the loads that drive various failure mechanisms; for example, limits on current density to manage electromigration, or minimum transistor length to manage hot carrier aging. These reliability design rules are defined by scientific investigations, often on discrete test structures (e.g., individual metal stripes for electromigration and oxide capacitors for time-dependent breakdown). The distribution of the time-to-failure for these structures then describes the capabilities of a given process's technology, and the reliability design rules are defined as a load condition that corresponds to a distribution or a point on a distribution that represents the desired intrinsic reliability.

2.2 Extrinsic Mechanisms

Extrinsic failures are due to process defects, rather than being inherent in the design and materials used. Extrinsic failures are dependent on both the defectivity associated with the semiconductor and packaging processes and the complexity of the design and manufacturing processes. It is assumed that the chip failure rate is a statistical representation of the interaction between these two attributes.

Extrinsic failures are typically managed through the applications of “best” practices used to improve process quality and yield, and the failure rate is normally evaluated through product level accelerated life tests. For well-managed designs, where the wearout has been addressed through the design rules, product reliability is controlled by the extrinsic failure rate.

3. INTRINSIC DEVICE SENSITIVITIES

The reliability of an IC is clearly dependent on the intrinsic reliability of the discrete transistors that comprise the IC. This chapter presents the various intrinsic failure mechanisms that affect the stability of a transistor and hence have an adverse effect on IC reliability.

3.1 Device Transconductance Sensitivities

The stability of a transistor is often expressed in terms of device 'transconductance' (g_m), defined as the ratio of change in output current to change in input voltage:

$$g_m = \partial I_C / \partial V_G \tag{3.1}$$

The modes by which g_m degrades varies for the major transistor types, including Metal Oxide Semiconductor Field Effect Transistors (MOSFET), Bipolar Junction Transistors (BJT), and Junction Field Effect Transistor (JFET). This section looks at the transconductance models for each transistor type and assesses, in general terms, the mechanisms by which it can degrade.

3.1.1 MOSFET Devices

Transconductance for a MOSFET transistor is given approximately as

$$g_m \approx \mu\, C_{OX}\, (V_G - V_T)\; W / L \tag{3.2}$$

where μ is the carrier mobility, C_{ox} is the oxide capacitance per unit area, V_G is the applied gate voltage, V_T is the threshold voltage, and W and L are the gate width and length, respectively [Streetman 1980]. Of these parameters, μ, C_{ox}, and V_T may be variables through the lifetime of the transistor.

Hot-carrier aging (Chapter 5) can affect all three of these factors, and hence can result in transconductance degradation [Chan 1984, Lee 1993, Yasuaky 1985]. Slow trapping, where carriers tunnel through a barrier formed by the energy band bending at the silicon-silicon dioxide interface, and are trapped in the gate dielectric, can also cause shifts in V_T [Lee 1993] and transconductance.

3.1.2 Bipolar Devices

For bipolar junction transistors, the transconductance is approximated by

$$g_m \approx \frac{q}{kT} \beta I_B \tag{3.3}$$

where q is the elementary charge, k is Boltzmann's constant, and T is temperature, and β is the base current amplification factor, and I_B is the base current.

The parameters which may change with time are β and I_B, which in turn, are dependent on diffusion lengths and carrier mobility. Being a bulk silicon device carrier mobility and diffusion lengths in BJTs are degraded only through phenomena with sufficiently high energy to disturb the silicon lattice, such as high energy (MeVs) neutrons or β-particle radiation [Billington 1961], or ESD/EOS effects. That is, transconductance degradation is not prevalent in BJTs except in special nuclear or space applications, or when the component is mishandled. Note however, that the parasitic MOSFET associated with every BJT, is susceptible to degradation, and can therefore affect BJT reliability, too.

3.1.3 Junction Field Effect Transistor

The equation used to describe the transconductance of a JFET transistor is

$$g_m = G_0 [1-(-V_G / V_P)^{1/2}] \tag{3.4}$$

where G_0 is the conductance of the channel and V_G is the gate voltage. The pinch-off voltage, V_P, is the characteristic threshold voltage [Streetman 1980].

In this equation the only parameter that may change over time is G_0, which is dependent on device geometry and carrier mobility and diffusion length. Thus, also being a bulk device, JFET, like BJT, is intrinsically stable, except in extreme situations. But, like BJT, JFETs are susceptible to parasitic surface effects.

3.2 Leakage Current Sensitivities

Junction and oxide leakage currents exist in any semiconductor device that is built up from p-n junctions and dielectric films. MOSFET devices have several types of leakage currents. The weak inversion current flows from the n+ drain to the n+ source in the n-channel MOS structure. This is negligible for long channel transistors with their high threshold voltage, but can be the dominant leakage mechanism for short channel transistors with low threshold voltage.

The drain-induced barrier lowering (DIBL) current is the channel surface current induced by lowering the threshold voltage, and there is a current induced in trench-isolated technologies when the gate width becomes too narrow. The electric fields around the narrow gate structures are modified such that threshold voltage decreases with width and this current increases. When the length between drain and source is short, punch through, an off-state leakage occurring deep in the channel, appears. It is called the reverse bias saturation current of the p-n junction that occurs in drain to substrate regions. This is the dominant IDDQ, low quiescent power supply current, which provides fast failure analysis by measuring current–voltage signatures and temporal characteristics in ICs with long channel transistors.

The gate-induced drain leakage (GIDL) occurs when a full V_{dd} bias voltage across the gate and drain terminals produces a large electric field at the oxide to drain interface. This intense field allows charge tunneling and the subsequent formation of hole electron pairs from the kinetic energy of the tunneled charge. Under normal operating conditions these leakage currents are small and are considered negligible. However, with elevated temperatures and voltages, leakage currents can increase and affect device performance and cause permanent aging.

Failures that are precipitated by excessive leakage currents include:

- Dielectric Failures – oxide leakage current tends to be phenomena with a positive feedback, eventually causing a runaway effect resulting in destruction of the dielectric film. Time-Dependent Dielectric Breakdown (TDDB), Chapter 6, is such a phenomenon, and is aggravated by any oxide leakage current, including that caused by hot carrier effect, or junction avalanching.

- Junction Failures – leakage currents in p-n junctions, especially under reverse bias, can also increase with time, and will eventually trigger a runaway second breakdown which will destroy the junction.

These failures can be precipitated either through exposure to long periods of use at or near the intrinsic capability limits of the semiconductor films and junctions (intrinsic aging), or through interactions with defects whose effect is to elevate *local* temperatures or electric fields (extrinsic aging), or by a combination of the two effects.

We consider that intrinsic leakage aging is an application variable; extrinsic leakage aging is a process variable. The control of the interaction of these two effects dictates managing IC sensitivity to process defects.

IDDQ is the most sensitive detection technique to detect defects that affect leakages currents. When ICs are designed and fabricated for low IDDQ, signature analysis of IDDQ versus VDD can be used for rapid identification of the root cause of defects and for improved corrective action.

3.3 Breakdown Issues

Under some conditions, semiconductor devices go into a 'breakdown' mode, described by a unique current–voltage (I–V) relationship, and typically corresponding to a situation where the current is no longer controlled by the applied voltage (Figure 3.1). Breakdown events are often runaway conditions that result in destruction of a device, and are typically triggered at some voltage that is above the normal operating range. There are several types of breakdown that can be induced:

- *Junction Breakdown* – given a sufficiently high reverse bias, a p-n junction will break down and conduct current. This is due to either avalanche multiplication of thermally generated carriers within a junction depletion layer, or to tunneling through the p-n junction barrier. The breakdown voltage BVpn is a function of the diffusion profiles, but for normal application can be maintained well above the operating range. Thus for typical CMOS (complementary MOSFET) devices the source/drain-to-well breakdown is of the order of 10 to 15V. For special cases, a diffusion profile can be tailored to either reduce the voltage to a few volts (as in Zener diodes), 3V and less for RF transistors, or to a few hundred volts (as in power transistors).

- *Punch-Through Breakdown* – with short channel MOS devices, an increasing applied voltage on the drain eventually causes a source-to-drain breakdown, called 'punch-through' (BVdss). This is due to the drain depletion layer spreading all the way to the source and causing barrier lowering of the source-to-substrate junction, resulting in carrier injection into the substrate. This is analogous to 'reach-through' breakdown in BJTs. Managing this problem is an issue in deep sub-micron devices, and usually calls for careful tailoring of the diffusion profiles so as to inhibit the drain depletion layer spreading across the channel.

- *Latch-Back Breakdown* – empirically, if the drain voltage is increased and the device is turned on (gate voltage is above threshold), a breakdown-like phenomenon takes place. This phenomenon appears to be a type of punch-through, but is actually caused by turning on parasitic bipolar devices associated with every MOS device (source-substrate-drain corresponds to n-p-n).

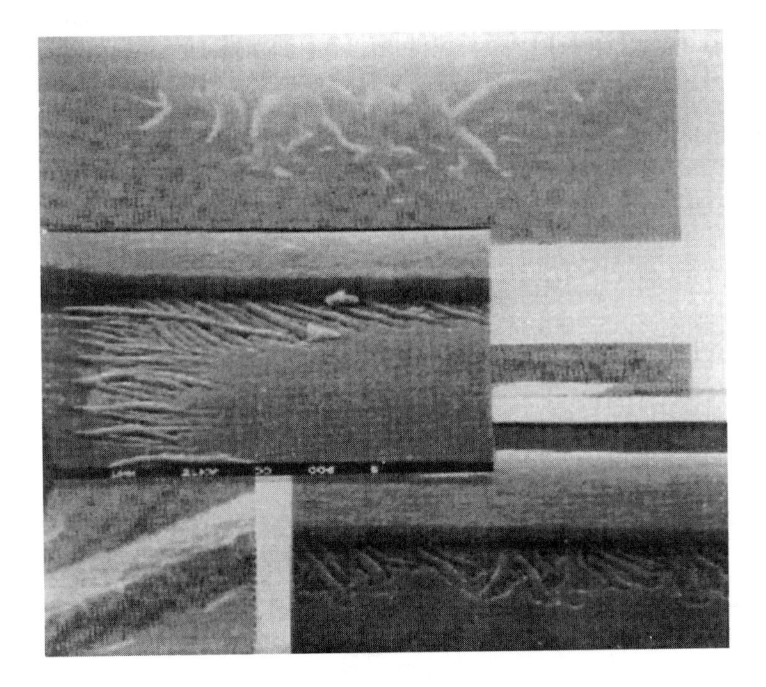

Figure 3-1 Some of the destruction induced by the breakdown

Impact ionization within the drain depletion layer of a short channel MOS device generates carriers which are injected into the substrate and act as a de-facto base current for the parasitic BJT. Latch-Back (BVdsg) current is really the current of the BJT. Depending on the architecture of a device, latch-back may be a positive feed-back phenomenon, and can therefore appear as latchup.

- *Secondary Breakdown* – this is a phenomenon that may be associated with other breakdown mechanisms. If a device is driven into breakdown through any of the above mechanisms, and the electronic field is sufficiently large, a thermal run-away mechanism may be triggered resulting in negative resistance and hence an uncontrolled rise in current.

All modes of breakdown are basically second-order phenomena that force device operation into some different region of the I–V characteristic. Breakdown phenomena typically drive a device into a region where minor changes in voltage result in major changes in current. Breakdown phenomena are often destructive, or very close to being destructive, unless there is some external current limiting mechanism. The destruction takes place when the breakdown-induced current is not uniformly distributed, but flows in a filament and causes localized excessive Joule heating. If the breakdown takes place close to the silicon surface, a degradation in the leakage or threshold characteristics may be induced due to trapping of energetic carriers that usually accompany breakdowns into the silicon-silicon dioxide interface states. The mechanisms may be either a time-zero issue or may result in a 'walking wounded' device that carries enhanced reliability risks with no externally visible or parametric evidence of a problem.

4. ELECTROMIGRATION

Electromigration is a failure mechanism which occurs in the interconnect metallization tracks. Electromigration is an intrinsic wearout mechanism, but due to interaction with defects, electromigration does not manifest itself until a circuit has been operated for months, or even years. Unless the problem is eliminated by design, deep submicron circuits will be failures waiting to occur. The phenomenon cannot be prevented by production testing.

4.1 Description of the Mechanism

Electromigration is a diffusion process which occurs when the current density carried by the interconnect tracks is sufficiently high to cause drift (diffusion) of the metal atoms in the direction of the electron flow (Figure 4.1). The number of ions that migrate in the 'electron wind', i.e., the ion flux density, is dependent on:

- The magnitude of forces that tend to hold the atoms in place, and thus the elemental nature of the conductor, crystal size, and grain boundary chemistry, and

- The magnitude of forces that tend to dislodge the atoms, including the electron current density, temperature, and mechanical stresses.

Polycrystalline aluminum films, which are typically used for interconnect in silicon chips, are most prone to electromigration due to the relatively loose binding of the aluminum atoms, especially at the surface of the individual crystal grains. Electromigration failures occur if a void, created at a point where the flux of the outgoing ions exceeds the incoming flux, becomes large enough to cause an open in the metal line. Conversely, a short to the adjacent or overhead metal runs is caused when aluminum atoms are piled up at a point where the incoming ion flux exceeds the outgoing flux.

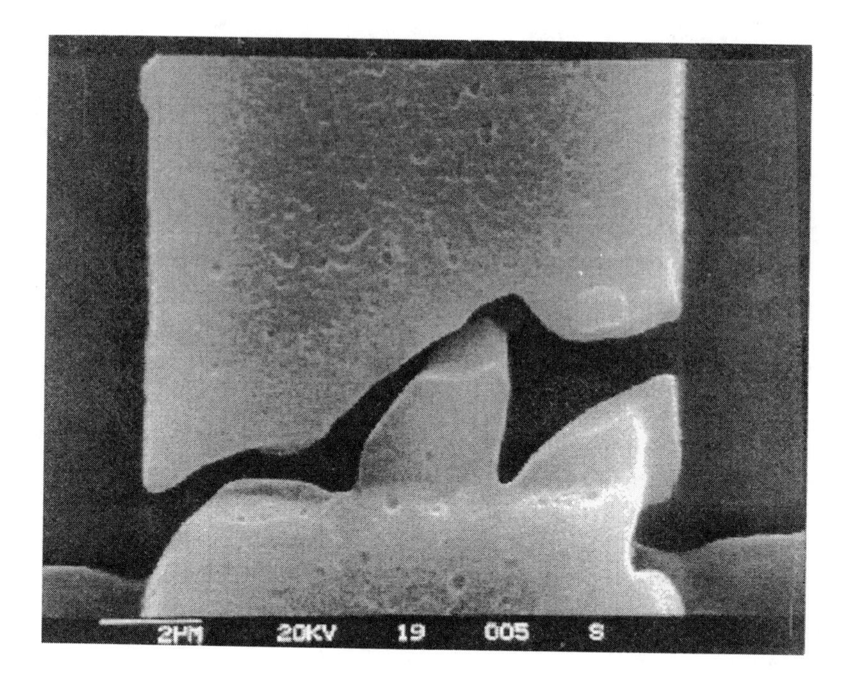

Figure 4-1 A typical open aluminum interconnect caused by electromigration.

Failures occur where there is a divergence in the ion flux, asymmetry in the ion flow, caused and/or aggravated by any factors, such as:

- Temperature gradients – where ion flow in or out of a hot region is higher than that in a cold region.

- Current density gradients – where ion flow in high electron current density regions is higher than in the low current density regions.

- Grain size gradients – where ion flow in a region with many grain boundaries is higher than in the region with few grain boundaries.

- Mechanical stress gradients – where ion flow in high stress regions is higher than in the relaxed regions.

- Ion flux barriers – where ion flow through a contact with a different material, such as titanium or titanium-tungsten, is inhibited.

All devices that use polycrystalline aluminum metallization are susceptible to electromigration failures. In the past, MOS devices with their inherently lower currents were less affected by this failure mechanism than BJT devices. However, with the continued down scaling of feature dimensions into deep sub-micron regions, and with operating frequencies reaching into many hundreds of MHz, reliability of the modern CMOS devices is critically impacted by electromigration, so that very concrete steps are required to manage this failure mechanism. Temperature as well as current density affect electromigration in a trace, but temperature may often be controlled after fabrication through the use of packages and heat sinks. Current density, on the other hand, must be controlled at the design stage. Preventive analysis, therefore, concentrates on locating areas of high current density. If dimensions are reduced for a fixed voltage, then internal electric fields will rise, increasing the reliability risk.

4.2 Modeling of the Mechanism

The drivers of the electromigration mechanism can be separated into environmental and material factors. The material factors include:

- Chemical composition – the content of the metallization system plays a dominant role, e.g., aluminum is prone to electromigration while copper or refractory metal conductors are not.

- Physical composition – the crystal grain size, i.e., the number of grain boundaries and the void density in the metallization also play a dominant role.

- Mechanical environment – the polarity (compressive or tensile) and magnitude of the mechanical stresses exerted on the metallization runs also impact electromigration.

The environmental factors include:

- Temperature – the mobility of aluminum ions is clearly a function of temperature and is typically described through an Arrhenius relationship.

- Electron current – the primary driver for the migration of aluminum is obviously the current density.

Electromigration may be modeled either as a mass transport (pile up of aluminum atoms) or a vacancy transport (precipitation of vacancies into larger voids) mechanism. In either case, the rate of material transport can be described as:

$$R = A\ J^n \exp(-Ea/kT) \qquad (4.1)$$

where R is the rate of material transport, J is the electron current density (in A/cm^2), kT is the Boltzmann constant (in eV) times temperature (in K), and A is an empirical constant that embodies the material factors. Ea is the Arrhenius activation energy (in eV), and n is an empirical current density exponent.

Time-to-failure or mean-time-to-failure is then inversely proportional to the mechanism rate. The acceleration factor, AF, that relates the ratio of the time-to-failure for different environmental conditions is given by

$$AF(T, J) = (\Delta\ J)^n \exp\ [Ea/k(\ 1/\Delta T)] \qquad (4.2)$$

where Δ J and ΔT are the difference in the current density and line temperature, respectively. The activation energy, Ea, is typically found to be between 0.5 eV and 1 eV depending on the material factors, and n is normally taken to be 2 by accounting for Joule heating.

The above models are applicable in a simple static environment, with constant currents and temperatures. The reality, especially for CMOS circuits, is that the currents experienced by the metal lines are switching and that the temperatures are not constant. Typically, the above relationships are then applied in real circuits using time-average currents and temperatures.

Further, there are no simple models that can predict time-to-failure characteristics of a metallization system based on the grain size distribution, chemical composition, or mechanical stresses. Whereas, the *qualitative* impact of these factors is understood, the *quantitative* relation between the material factors and electromigration is typically derived empirically.

4.3 How to Detect/ Test

The complex nature of the electromigration mechanism requires some empirical calibration. This is normally accomplished by performing accelerated life tests on simple discrete test structures, such as single metal lines, single lines crossing over underlying topography, or chains of metal links at different metallization levels. Life tests are performed at elevated temperatures (typically > 180°C where polycrystalline aluminum tends to restructure), and with high current densities (typically > 10^7 A/cm^2 to induce accelerated Joule heating effects), and an assumed distribution (typically log-normal) of cumulative percentage failures vs. time. Ideally, this type of test should be performed as a design-of-experiment, on a statistically significant sample of every type of feature that is used (e.g., for every metal level, for vias and for contacts), and should be performed under different current and temperature conditions in order to extract the value of Ea and n. It is common practice however to abbreviate the sampling and diversity of the test structures and to use the industry standard value of Ea and n. Normally, the material factors will be adjusted through process changes until the mean and standard deviation of the log-normal distribution are such that cumulative percentage failures of about 0.1% correspond to over 100,000 hours under normal in-use conditions.

4.4 How to Manage

Electromigration can be managed by controlling the material factors, the enviromental factors, or both. The material factors are controlled by the process technology and the manufacturing practices. Current state-of-the-art practices for managing the material factors of electromigration include:

- The addition of small amounts of copper or tungsten (~ 0.5 wt. %) dopants to the aluminum can greatly inhibit electromigration by 'passivating' the grain boundaries through binding of the dangling aluminum atoms at the crystallite surfaces. Annealing in a hydrogen environment is also known to bind the aluminum atoms. However, excessive he dopants can have an adverse affect on other failure mechanisms, such as corrosion for copper or stress migration for titanium.

- The addition of a thin layer of a refractory metal, such as tungsten or titanium-tungsten, under or on top of the aluminum metallization run, is desirable, not only as a source of beneficial dopants, but also as a shunting layer that is immune to electromigration. Thus, void generation will not necessarily lead to a catastrophic circuit failure (open line). However, shunting layers tend to increase line resistance and complicate the electromigration dynamics in the vias and contacts.

- The use of process techniques that control grain size (typically ~0.25 to 1 μ), including the deposition temperature and the number of thermal cycles (>180°C) have a vital impact. However, managing the thermal history is challenging, especially in multi-level metallization systems, where successive process steps demand use of elevated temperatures.

- The control of the mechanical stresses, especially associated with the nature and process of the surrounding insulators, is essential. Mechanical stresses can induce voiding or hillocking in the metal lines, which then aggravate the electromigration mechanism. This is challenging because the common conductor material is aluminum with a thermal coefficient of expansion (TCE) of the order of 25 ppm, and the common insulator material is silicone glass with a TCE of the order of 3 ppm.

The environmental factors, temperature and current density, are controlled by the chip design and the application environment, and for a given silicon process technology, are the only variables that can be controlled.

- Current density is often controlled through a 'design rule' limit that corresponds to the desired reliability level. For most metallization systems this current density limit is of the order of 100,000 A/cm^2 at about 55 °C. The design rule limits are extracted empirically from the log-normal failure distributions derived in the accelerated life tests on discrete test structures. With high performance deep sub-micron products, this type of current limit is a real limitation in design and needs to be managed carefully.

- Temperature is controlled by the power dissipation of the chip, the package, and by the system cooling environment. Here too, besides reliability, there are cost and performance trade-offs that need to be carefully managed.

5. HOT CARRIER AGING

Hot carrier aging is defined as the degradation of intrinsic MOS device characteristics due to the trapping of charge in the gate dielectric. 'Hot carriers' are those carriers (electrons or holes) that are not in 'thermal equilibrium' with the rest of the semiconductor crystal lattice; i.e., the carriers whose energy does not correspond to that of the conduction and valence band edges. This situation arises when the electric field seen by the carriers causes the carrier acceleration between the collisions with the lattice i.e., within a mean free path, to be sufficiently high to change their average energy [Jagdeep, 1992].

High energy hot carriers can cause a number of effects within a MOS device, including a change in device threshold voltage and transconductance. Degradation occurs without interactions with any defects or imperfections, and is a wearout phenomenon. Hence hot carrier aging phenomenon can only be designed out, and cannot be screened out.

Hot carrier aging occurs in both the MOS and the BJTs, but is inherent in MOSFETs, while the effect is more circuitous in BJTs. The fundamental driver of the mechanism is the electric field, and the degradation rate increases as device dimensions are decreased. This mechanism is a fundamental issue in modern sub-micron devices.

5.1 Description Of The Mechanism

In metal oxide semiconductor field effect transistor (MOSFET) devices, electric fields are utilized to control the flow of electrons or holes through a device. Depending on the polarity of the source and drain regions and the threshold voltage, these devices are characterized as either NMOS or PMOS transistors and are operated in either enhancement or depletion mode, respectively. In any case, for digital applications the MOSFET operates as a switch where, the gate-to-source potential (Vgs) determines if the switch is open or closed, and the drain-to-source (Vds) potential determines the magnitude of the current flowing through the switch.

One of the consequences of 'constant voltage scaling', where the device dimensions are decreased in every successive generation of technology while the operating voltage is maintained at some constant value (e.g., Vdd =5.0 V), is that the intensity of the electric fields within the device is increased. Specifically, the source-to-drain electric field, and especially the electronic field around the drain region, has been increased in every successive generation of technology. Consequently, in modern sub-micron devices this electronic field is sufficiently high to cause carrier heating effects, including:

- Impact ionization and avalanche multiplication in the drain region, resulting in increased drain current and a substrate current, and,
- Injection of carriers into the gate dielectric region, resulting in gate current.

The hot carriers generated by the electronic field may become trapped either in the bulk of the gate dielectric, or in the interface region between the dielectric and the semiconductor. In either case, the trapped carriers perturb the normal distribution of the electronic fields within a device, which in turn results in a shift of the threshold voltage and the device on-resistance (transconductance). Over time, as more and more carriers are trapped, degradation of the device characteristics can be significant, and can affect the circuit performance [Matsumoto 1981, Cemal 1994].

Hot carrier aging rate is controlled by the net interaction of three independent variables:

- Hot Carrier Generation Rate - the number and energy of hot carriers generated by the electronic fields per unit time. This is controlled by the the magnitude of the peak electronic fields, and hence is dependent on the applied voltage, device geometries, and the junction diffusion profiles. In addition, carrier mean free path, controlled by temperature and material properties, is a primary variable.

- Carrier Trapping Rate - the number of hot carriers per unit time that become trapped in either the bulk gate dielectric or in the dielectric-to-semiconductor interface. This is controlled by the nature of the materials and is dependent on the material properties and the process steps used to build the device.

- Device Aging Rate - the impact on device operation per unit of trapped charge. This is controlled by the architecture of the device as well as the application parameters.

Hot carrier aging is an inverse function of temperature [Hsu and Chiu 1984, Cemal 1994], due to the reduction of carrier mobility (mean free path) with increasing temperature, and due to the thermal re-emission of trapped carriers in the silicon oxide [Cemal 1994]. The aging effect is a very strong function of the applied voltages, since these control the internal electronic fields. Furthermore, hot carrier aging is a complex function of the relation between Vgs and Vds, as this relationship drives the distribution of the internal peak electronic fields. Finally, the magnitude of the substrate current (I_{sub}), is a direct measure of the impact ionization in the drain region, and is hence used as a measure of carrier heating.

Typically, device lifetime has been defined as the time required to cause a 10% reduction in the device transconductance (g_m), and/or 0.1V shift in the threshold voltage (Vth) [Cemal 1994].

5.2 Modeling of the Mechanism

Hot carrier aging can be modeled as a function of the ratio of the substrate to drain currents. According to Hu [1985] a model can be expressed in the form:

$$\tau = H\, I_{sub}^{-2.9}\, I_d^{1.9}\, \Delta V_t^{1.5}\, W \tag{5.1}$$

where τ is the device lifetime defined as the stress time required to achieve a threshold voltage shift of 100 mV, I_{sub} is substrate current, I_d is drain current, Vth is the threshold voltage, W is the channel width, and H is the dielectric technology scaling factor.

This relationship can be derived from the first principles founded in the fundamental device physics, and is then calibrated using empirical data. The factors that are used in the derivation include the carrier mean free path, the barrier height, and the critical impact ionization energy for silicon. I_{sub} is a measure of the hot carrier generation rate. Broadly, H encompasses the hot carrier trapping rate, and ΔV_t includes the device aging rate factors.

Figure 5.1 shows the characteristic device lifetime as a function of Vdd and temperature. Figure 5.2 shows a plot of device lifetime defined as the time to achieve a 10 mV change in Vth versus Isub and indicates the power law relationship with a slope of 2.9. Figure 5.3 shows the lifetime versus Isub for various device technologies, indicating that while the slope remains constant at 2.9 the value of h for a fixed Isub is dependent on oxide/interface quality or technology.

5.3 Detection of Hot Carrier Aging

Hot carrier *generation* can be detected either by measuring substrate current (I_{sub}) or by measurements of light emission. As indicated, substrate current is a direct measure of the impact ionization process, and in effect, counts the number of holes (electron-hole pairs) generated in the avalanching process. Light emission measurements can evaluate the number and energy of photons emitted when the hot carriers recombine, and should therefore be a most direct method. However, since substrate current is particularly easy to measure, and photon detection is particularly hard to quantify, the industry standard techniques rely on I_{sub} measurements.

Hot carrier *aging*, however, includes the charge trapping phenomena, and the effects of the trapped charge on the device, over and above just the hot carrier generation rate. Thus, measurement of the hot carrier generation is only a qualitative metric of the aging rate. Quantitative measures of aging are performed through accelerated tests on discrete transistors, usually at elevated Vdd, and time-to-failure is measured directly. The degradation rate is then modeled to predict life under the in-use conditions. As is often the case with accelerated tests on discrete test structures, care must be taken to ensure that all types of transistors are characterized and that the translation from accelerated conditions to in-use conditions encompasses all relevant factors (I_{sub}, Vdd, Vds/Vgs, T, operating frequency and switching slew rate).

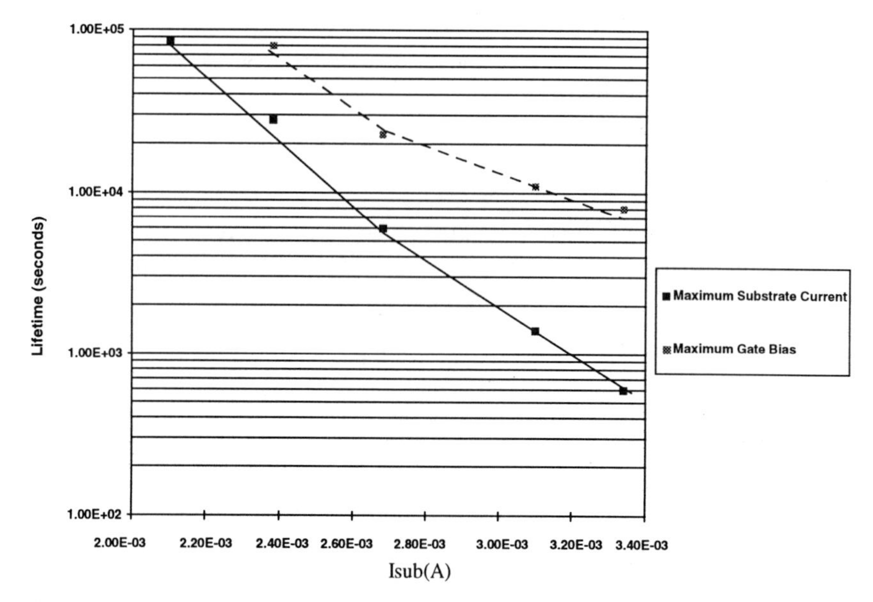

Figure 5-1 Characteristic device lifetime as a function of Vdd and temperature [Solid State Electronics, 1994].

In addition, silicon process and material factors that influence the trap density in gate dielectric or the dielectric-to-semiconductor interface, must be encompassed in the tests. Some of these process and material factors are quite subtle, so that any form of abbreviated testing carries some risk, and must be designed carefully.

5.4 Avoidance of Hot Carrier Aging

The effect of hot carrier aging can be managed by addressing one or all of the three factors that control the net aging rate:

- Control of hot carrier generation rate: the most direct method for reducing the hot carrier genertion rate is to reduce the applied voltages. The industry, however, has traditionally been very reluctant to take this step because of the system level needed to maintain some standard Vdd value. An alternative approach has been to reduce the *peak* electronic fields within a device by suitable tailoring of the junction diffusion profiles. Use of 'Low Doped Drain' (LDD) structures, which tend to smear the drain electronic field distribution, has therefore become pervasive in all sub-micron technologies.

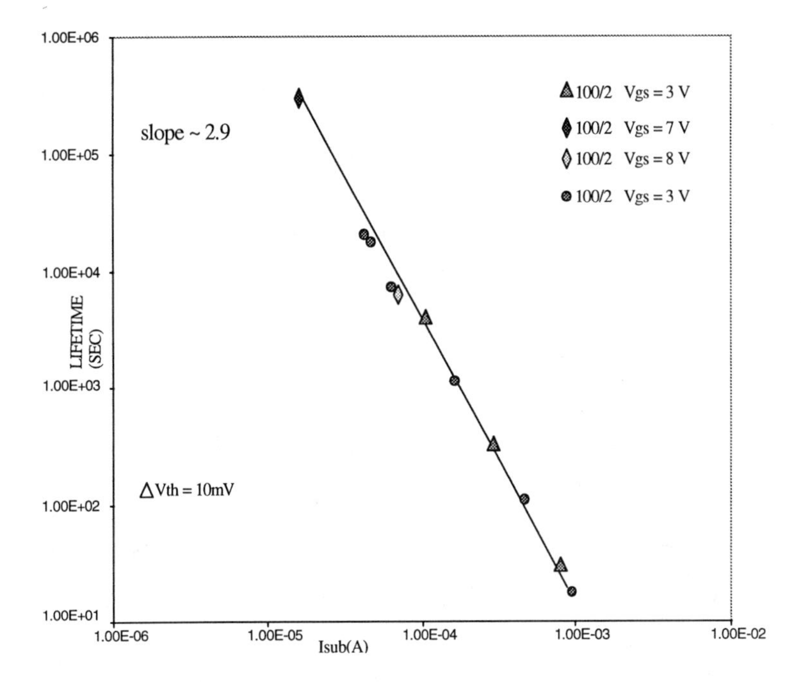

Figure 5-2 A plot of device lifetime τ, defined as the time to achieve a 10 mV change in V_{th}, versus I_{sub} and indicates the power law relationship with a slope, 2.9.

- Control of carrier trapping rate: as indicated the trapping rate is a function of the material and process variables. Careful control of the processing temperature profiles, material impurity content, and sequencing of annealing cycles is required to minimize the trap density in gate dielectric, and the industry has evolved many process recipes to achieve this. In addition, use of novel materials, such as silicon oxide, silicon nitride mixtures, can reduce trap densities.

- Control of device aging rate: this can be controlled through either the architecture of the device or application conditions. It is possible to devise a device which can absorb a lot of trapped charge with very little shift in the device parameters by ensuring that the peak electronic fields do not coincide with the critical regions in the device. In addition, use of circuit design practices that either ensure that a device spends little time in the bias region where the carrier generation rate is high, or ensure that the device is operated in the bias region where degradation of the transistor parameters is not critical, is becoming pervasive. This is typically achieved by implementing a set of 'design rules' that describe the safe operating regions for a given device type.

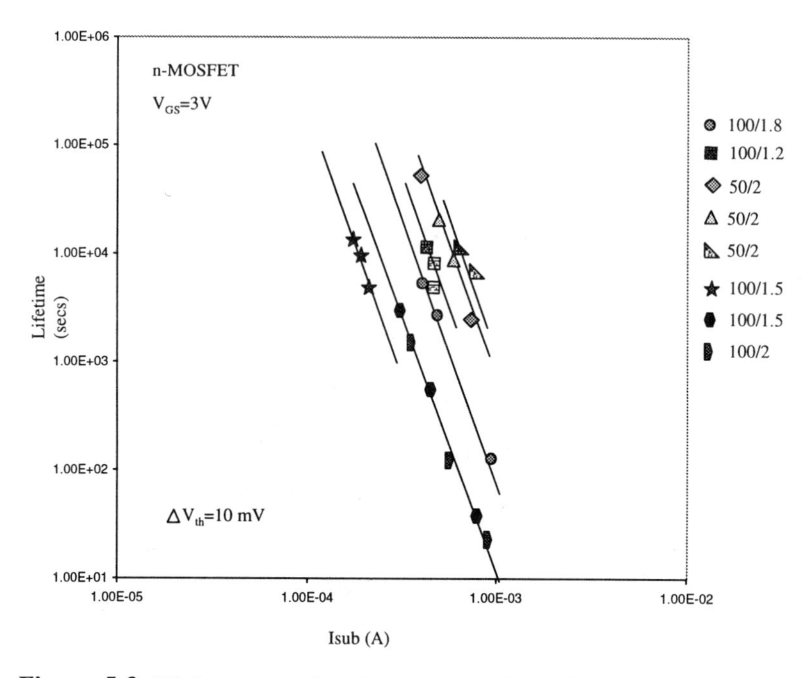

Figure 5-3 Lifetime versus I_{sub} for various device technologies, indicating that while the slope remains constant at 2.9 the value of h for a fixed I_{sub} is dependent on oxide/interface quality or technology [Hu, 1985].

As is often the case, optimizing the device and/or the process for hot carrier aging may have an adverse impact on some other variable. Thus, typically, at the device architecture level, the measures that reduce hot carrier effects tend to also reduce device performance. Some of the process measures that have a favorable effect on hot carrier aging may have an unfavorable effect on ESD immunity.

6. TIME-DEPENDENT DIELECTRIC BREAKDOWN

Dielectric Breakdown is a failure mechanism which occurs in thin dielectric films, such as is used for the gate in a MOS device. The dielectric strength of an oxide layer is often expressed in terms of the electric field at which has its insulating properties and the insulator is irreversibly damaged. With a sufficiently high applied electric field, and after continued exposure to a *constant* applied voltage, dielectric films will breakdown. This phenomenon, called 'Time-Dependent Dielectric Breakdown,' is a true wearout mechanism, but due to the interaction with defects in the dielectric film, it may be observed in all three regions of the bathtub curve.

6.1 Description of the Mechanism

Dielectric Breakdown occurs in silicon dioxide films when the local electric field is sufficiently high to cause avalanche multiplication in the dielectric. That is, silicon dioxide is a dielectric with a property analogous to a 'band gap' of the order of ~9 eV. When the applied electronic field is sufficient, normally around 10 mV/cm to 15 mV/cm, the electrons that tunnel in from the cathode cause impact ionization within the dielectric and the consequent current destroys the material through sheer Joule heating. In silicon dioxide this is a run-away phenomenon, because the holes, either tunneled in from the anode or produced by the impact ionization, are trapped (hole mobility in silicon dioxide is 10^{-5} $cm^2/Vsec$), creating a positive charge cloud that further enhances the electronic field. The electrons (mobility ~30 $cm^2/Vsec$) hence are multiplied and cause the destruction. Clearly, the phenomenon is triggered when the electronic field reaches the critical value (E_{crit}) required to cause carrier multiplication. Due to the minute imperfections in the film, the electronic field is normally not perfectly uniformly distributed, so that this phenomenon tends to take place in a local region, and the current flow and the consequent destruction occur in a filament in the film, in effect a localized explosion occurs (see Figure 6.1). This results in a total denaturing of the region and typically causes a short between the gate electrode and source/drain or substrate electrodes. If the film is close to perfectly uniform, the breakdown phenomenon takes place, but the local current densities are not excessive, and the film is not physically destroyed. However, this non-destructive breakdown typically results in films that are filled with trapped charge, and consequently device threshold voltages are shifted. In either scenario, the device ceases to function. The use of the Q_{BD} concept in reliability predictions strongly depends on the behavior of Q_{BD} as a function of stress current density. This can be used for intrinsic reliability estimates of gate oxide and the switching degradation.

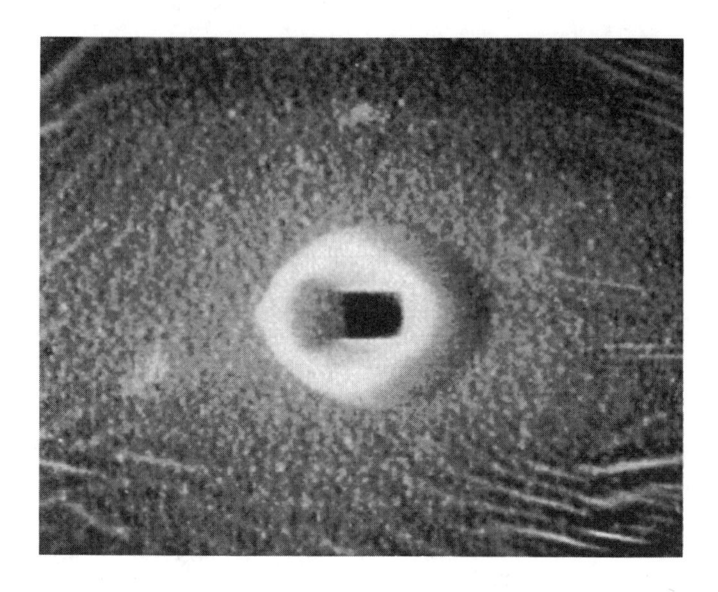

Figure 6-1 An image of a site of dielectric breakdown.

A constancy of Q_{BD} would allow a simple estimate of time to breakdown use via the equation, $t_{BD} = Q_{BD} / J_u$, where J_u is the current density under stated conditions. However, it indicates that the values of Q_{BD} at these high injection levels are not relevant for reliability prediction. It is the phenomenon that Q_{BD} is not constant for oxide layers as thin as 10 nm, but that it increases with decreasing oxide voltage.

There is no apparent discontinuity in breakdown behavior as the film thickness goes below 100 A; gate oxides of the order of 50-70 Å thick are the norm at 0.25 µ technology level, and suffer breakdown at about 7 to 10 V.

Since the mean free path in silicon dioxide is of the order of 50Å, carriers basically fly through very thin dielectrics with no collisions, indicating that the impact ionization mechanism is not responsible for the breakdown in these films. There has been some speculation that the breakdown in very thin dielectrics is due to micro-plasma events within the film, but this model is not as yet generally accepted throughout the industry.

Note also that the tunneling distance for the silicon-silicon dioxide interface is also of the order of 50 Å, indicating that films thinner than this basically do not act as dielectrics since carriers can tunnel directly through them. Consequently, the industry is experimenting with various film compositions for technologies scaled beyond ~0.25 µ range.

The dielectric breakdown phenomena outlined above typically occur when the applied voltage is increased. With time-dependent breakdown, the same phenomenon occurs after a prolonged exposure to a *constant* applied voltage. Clearly, breakdown occurs due to some redistribution of electronic fields, and the 'time-dependent' portion of the phenomenon is a function of the time required to reach the critical electronic field for the breakdown (E_{crit}). That is, with TDDB, an outline of the sequence of events is:

1. An oxide film is stressed at some applied voltage.
2. This results in an electronic field that can be represented as a distribution of values, controlled by the uniformity of the film thickness and composition.
3. The applied electronic field results in gate current - due to electrons tunneled in from the cathode.
4. Some of the electrons are trapped in silicon dioxide.
5. In addition, holes are injected from the anode and are trapped, and any mobile charged contaminants are induced to drift into the electronic field.
6. The trapped charge causes a redistribution of the local electronic field.
7. The redistribution of the electronic fields eventually causes some local value to reach critical value $E_{crit.}$
8. Dielectric breakdown and the associated runaway are triggered.

Thus, the TDDB time-to-failure is dependent on the rate of redistribution of the electronic field values, i.e., the time taken to reach E_{crit} value. That is, the breakdown voltage (BV_{ox} corresponding to E_{crit}) and the TDDB time-to-failure are in fact independent variables, although, clearly, high E_{crit} usually translates to good TDDB characteristics. Presence of a defect within a dielectric film can act to focus and enhance local electronic fields, thereby reducing the apparent BV_{ox}, but, for a given material, not the E_{crit} values. A statistical distribution of random defects then tends to produce a distribution of BV_{ox} and TDDB time-to-failures. In addition, with modern 'dry processing' the dielectrics are exposed to high voltage plasmas which can induce dielectric breakdown during wafer processing. That is, the oxides can be either destroyed or worn out during the manufacturing process, resulting in an altered BV_{ox} and TDDB distributions. This effect is design dependent, as the amount of charge absorbed by the dielectric film during processing is related to the size, shape and connectivity of the features connected to it, such as the length of metal runs. Consequently, this is often referred to as the 'antenna effect.'

All devices that use silicon dioxide based dielectrics are susceptible to the TDDB phenomena. The insulator films used as the gate dielectric within a MOS device, or as a capacitor dielectric in dual level poly technologies, are typically orders of magnitude thinner than the films used as intermetal level dielectrics, and hence experience much higher electronic fields, and are therefore the features of real concern. Modern BJTs sometimes use thin dielectric spacers which can also be susceptible to TDDB.

6.2 Modeling of the Mechanism

With modern technologies and very high quality dielectric films, the drivers of the TDDB mechanism, described above, can be separated into environmental, material and interacting factors:

- Material Factors – include film properties that control parameters such as barrier heights, trap density, carrier mobility, interface state density and mobile contaminants.

- Environmental Factors – include parameters such as the applied voltage and temperature and control tunneling current, trapping and de-trapping rates and carrier mobilities.

- Interaction Factors – include parameters such as the defect density and size distribution, the antenna nature and size, thermal process and plasma voltage history.

The material factors are controlled by the deposition processes, and are expressed through the 'Fowler-Nordheim' relationship, and the 'Shottky-Reed' trapping statistics. The Fowler-Nordheim relationship describes the probability of carrier tunneling from cathode into the dielectric, and encompasses factors such as the barrier height, carrier effective mass and mobility, and is described by the I-V characteristic of the dielectric. The Shottky-Reed statistics encompass factors such as the trap density and cross section, and trap energy depth which are not readily described through some direct empirical measurements. In both of these relationships the magnitude of the local electronic field is a dominant factor, and hence the distribution and nature of defects have an overriding impact.

The environmental factors, i.e., the voltage and temperature, are controlled by the application environment and are the normal variables in the models.

Due to the complex and statistical nature of the interaction factors, the standard practice is to describe TDDB through an empirically derived acceleration factor that relates time-to-failure under some elevated stress condition to time-to-failure under normal in-use conditions. This acceleration factor is normally expressed as:

$$AF(T, E) = A^{\cdot} \exp(\beta \, \Delta E) \exp[Ea/(k\Delta T)] \qquad (6.1)$$

where AF is the acceleration factor, Δ E and Δ T are the difference in the applied average electronic field and temperature, respectively, β and Ea are material factor-driven constants, and A is an empirical factor that addresses the interaction. For a constant applied electronic field (Δ E = 0) activation energy (Ea) is typically found to be low – of the order of 0.3 eV. For a constant temperature (Δ T = 0) the field driven acceleration factor (β) corresponds to AF values ranging between ~50/mV/cm to 10^7/mV/cm, depending on film thickness and quality, and the experimental protocol. That is to say, currently there is no industry standard value of β. In fact, there is a school of thought that indicates that the acceleration factor is related to $\Delta(1/E)$ rather than ΔE. Furthermore, temperature also affects the electronic field acceleration factor, further compounding the complexity of the relationship.

The above model is applicable in a simple static environment, with constant electronic field and temperature, and does not account for any relaxation that can take place with switching stress.

6.3 How to Detect/ Test

As indicated, due to the complex nature of the TDDB mechanism and the statistical interaction with defects, empirical calibration of the reliability of a dielectric system is required. This is normally done by performing accelerated life tests on simple discrete test structures, such as simple capacitors, capacitors with large area or with long edges. Life tests are typically performed at elevated voltages (typically corresponding to electronic fields > 7 mV/cm to 10 mV/cm), and sometimes at elevated temperatures (typically > 150 °C), and a distribution (typically log-normal) of cumulative % failures vs. time is derived.

Ideally, this type of test needs to be performed on a statistically significant sample of every type of feature that is used (e.g., dielectric for p-MOS and n-MOS devices, and with and without large antennas), and should be performed under different voltage and temperature conditions, in order to extract the value of Ea and β. It is common practice however to abbreviate the sampling and diversity of the test structures and to use industry standard values of Ea and β, at best, to empirically extract β only. Normally, the material factors will be adjusted through process changes until the MTTF and σ of the log-normal distribution are such that cumulative % failures of < 0.1% correspond to >100,000 hours under normal in-use conditions.

6.4 How to Avoid and Manage

TDDB is managed through the material, environmental, or interaction factors, or all of these, by controlling the process or the design and application variables. The process variables used to manage TDDB include:

- Careful control of the oxidation conditions, such as the temperature and temperature ramp rates, to limit the stresses, trap densities and uniformities.

- Addition of controlled amounts of dopants, such as Cl or N, and H, to passivate the interface and dangling bonds.

- Use of a sandwich of a grown and a deposited dielectric to minimize pin-hole density while maintaining control of the interfaces.

- Careful control of the plasma and ion implant processes including voltages, polarities, currents and grounding schemes to manage the antenna effect.

- Careful control of the cleaning processes, especially prior to the growth of the active oxides, typically including a sacrificial oxidation cycle to manage the defects and interfaces.

- Defect control, defect control, defect control, defect control, defect control.

The design and application variables used to manage TDDB include:

- Control of the applied voltages and temperatures, and transients to ensure that the oxide is not being overstressed.

- Control of the layout to manage the antenna effect. This is normally implemented through definition and verification of antenna design rules that limit the area of conductors connected to the dielectric.

7. MECHANICAL STRESS-INDUCED MIGRATION

Stress migration is a failure mechanism which occurs in the interconnect metallization tracks and results in failure modes similar to electromigration, line opens and/or shorts. Stress migration is caused by the mechanical stress experienced by the metal lines versus the electron currents that cause electromigration. Stress migration is an intrinsic wearout mechanism, but due to interaction with other mechanisms, such as electromigration, it is observed in all three regions of the bathtub curve.

7.1 Description of the Mechanism

Stress migration occurs when the mechanical stress on the aluminum lines is sufficiently large to cause aluminum ions to migrate along stress gradients. Stress migration is a basic creep type of phenomenon, where, given the right magnitude and polarity of stress, and adequate time and temperature, aluminum material 'flows' in order to relieve the stress. Thus stress migration is fundamentally a diffusion process with a superimposed stress gradient-induced drift. The phenomenon is often interpreted as a vacancy transport mechanism, where the vacancies drift in the stress field and precipitate into voids, rather than a mass transport mechanism where aluminum ions drift. In either case, the mechanism is a result of an equilibrium between:

- The forces that tend to control the 'fluidity' of the aluminum line, driven by factors such as temperature, material composition, and line geometry.

- The forces that tend to control the superimposed ion (vacancy) drift caused by mechanical stress gradients, by factors such as the thermal coefficient of expansion (tce) of the metal and the surrounding glass and thermal history associated with the process.

Thus, polycrystalline aluminum films, which are typically used for interconnection in silicon chips, are prone to stress migration due to the relatively loose binding of aluminum ions (high mobility and density of vacancies) i.e., aluminum is the softest material in the superstructure system and is most prone to creep, and because the stress levels, induced mostly by the mismatch in the tce of aluminum and surrounding dielectric, can be very high.

Stress migration failures occur if a void, created when the metal line is under tensile stress, at a point where the vacancies precipitate, becomes large enough to cause an open in the metal line. Conversely, a short to the adjacent or overhead metal runs is caused when the metal line is under compressive stress and aluminum hillocks become large enough to bridge the dielectric gap. Stress migration can interact with electromigration; aluminum creep can create a void in the metal line which constricts the electron current flow, so that the remaining metal line width is opened due to electromigration. Similarly, a crack in dielectric glass causes a significant stress gradient so that stress migration forces the aluminum to fill the crack and create a short. And so on.

Typically, stress migration tends to create big voids with shapes that mirror crystal grains, following extended exposure to time and temperature, or following slow excursions into temperature regions that cause restructuring of aluminum lines (see Figure 7.1). In contrast, electromigration tends to result in slit- or wedge-shaped opens in the lines, following extended exposure to temperature and current density; and shear can cause cracks in metal lines typically following sudden excursions into low temperature regions.

All devices that use polycrystalline aluminum metallization are susceptible to stress migration failures. However, factors that increase this problem are:

- Reduction in the metal line widths – wide metal lines are 'spongy' enough to absorb the stress gradients without catastrophic creep failures. Long skinny lines are at highest risk of stress migration failures.

- Use of multi-level metallization systems. The multi-level metal process typically calls for temperature excursions to some elevated temperature, say a few hundred °C, for deposition of each dielectric and metallization level. These multiple temperature excursions, coupled with the fact that the tce of aluminum is of the order of 25 ppm/°C, while the tce of typical silicon-oxygen based dielectrics is of the order of 3 ppm/°C, inducing stress levels of the order of 10^9 dynes/cm, can cause extensive voiding and/or hillocking in aluminum lines.

7.2 Modeling of the Mechanism

The drivers of the stress migration mechanism described above can be separated into environmental and material factors. The material factors include:

- Chemical composition – the content of the metallization system plays a dominant role, e.g., aluminum is prone to stress migration while refractory metals are not.

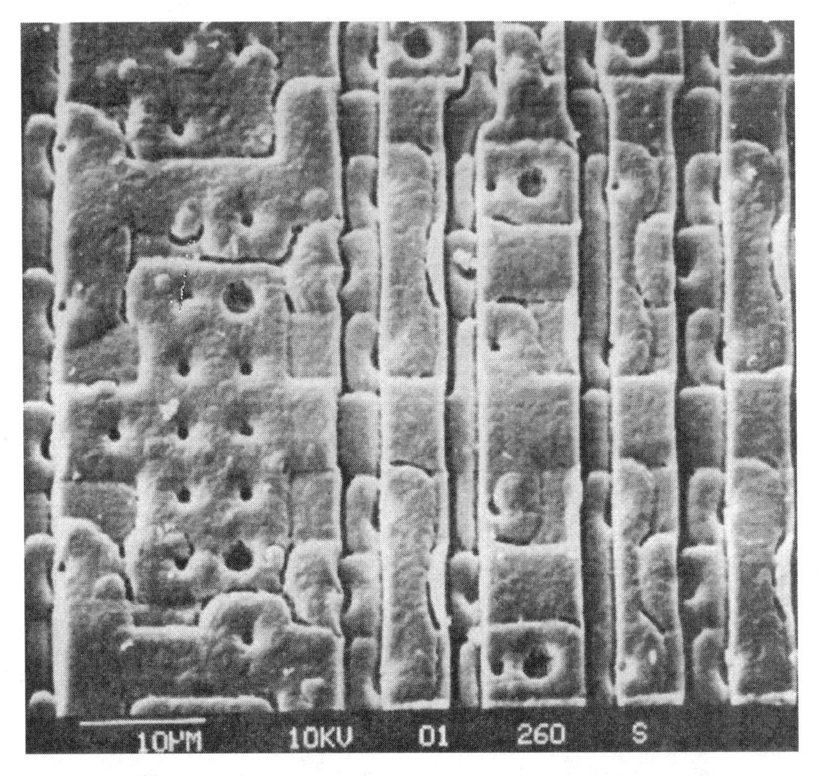

Figure 7-1 An open aluminum interconnect caused by stress migration

- Physical composition – the crystal grain size, i.e., the number of grain boundaries and the void density in the metallization both play a role.

- Mechanical stress environment – the polarity (compressive or tensile) and magnitude of the mechanical stresses, induced by the differences in tce exerted on the metallization runs drive the phenomenon.

The models that describe the impact of all material factors on stress migration are, as with electromigration, typically expressed in terms that cannot be directly measured or related to failure distributions, and are hence of limited practical use. The fundamental mechanical stress, induced by the TCES differences, is typically evaluated by measuring the bow in the wafers using laser interferometry-based tools. This technique can only be used with unpatterned wafers and is hence typically used with test wafers to assess the average strain in each individually deposited film, relative to the silicon substrate. That is, this technique is not compatible with evaluating the real stress induced in patterned multi-level die, and is therefore a qualitative indicator used to control the process. Data from this technique could be used to calibrate a finite element model, but this is a complex effort that does not address many events that could take place at the interfaces.

The environmental factors include:

- Temperature – the mobility of aluminum ions is clearly a function of temperature, and is typically described through an Arrhenius relationship.

- Thermal history – the number of thermal cycles that the system has experienced, especially with temperature excursions beyond the plastic point of aluminum, where the film restructures and relieves the stresses, is also critical.

Consequently, today there are no straightforward models that relate quantitatively time-to-failure and/or failure probability for stress migration. General practice is to tune the metal and glass deposition processes to minimize stress voiding, and to encompass any residual effects within the electromigration model. That is, by deriving the electromigration failure distributions on test structures that are sensitive to stress effects, the stress migration impact on reliability is encompassed within the electromigration distributions.

Note, however, that whereas electromigration is driven by time-average currents, the dominant design factor that affects stress migration is sheer power density in the metal line, related to the rms value of the current.

7.3 How to Detect/Test

Stress migration is typically detected either directly, through inspections, or indirectly, through electromigration tests. With multi-level metallization systems, visual or SEM inspections are limited, as the effect is usually most pronounced in the first level of metal because it experiences the greatest stress and the highest number of thermal cycles, which is obscured by the subsequent levels of metal and glass. Consequently, visual inspection calls for de-processing, which is often invasive, because the de-processing procedure itself can affect the observation, and is always time consuming. There are sophisticated tools, based on imaging a thermal wave created by a laser pulse, that allow detection of voids in a fully processed die, but these are costly and allow inspection of only one line at a time. Consequently, standard practices used in the industry call for use of specific test structures that are either evaluated in conjunction with electromigration tests, or are tested separately following exposure to creep conducive types of environment, such as slow thermal cycling through temperatures in excess of ~180°C.

7.4 How to Manage

As with electromigration, stress migration can be managed by controlling the material factors or the enviromental factors, or both. The material factors are controlled by the process technology and the manufacturing practices. Current state-of-the-art practices for managing the material factors include:

- Addition of small amounts of copper or tungsten dopants to inhibit the fluidity of aluminum.

- Addition of a thin layer of a refractory metal, such as tungsten or titanium-tungsten, under and/or on top of the aluminum metallization run, is desirable not only as a shunting layer, but also as a stress relief film with a tce that is between that of aluminum and silicon-oxygen based dielectrics.

- Use of process techniques that control grain size and vacancy density, such as deposition of aluminum at temperatures close to the ones used for deposition of glass, or deposition of aluminum and a refractory metal without breaking vacuum.

- Control of the mechanical stresses, by tailoring the dielectric deposition processes to minimize the effects. The TCE and stress of the dielectric are controlled by the film stoichiometry (silicon vs. O content) and dopant content (e.g., H or N), as well as by the deposition temperature. In addition, use of two films in the dielectric layer, one compressive and one tensile, is quite common.

The environmental factors, temperature and thermal cycling, are controlled by the component design and the application environment, and for a given silicon process technology, are the only variables that can be controlled.

- Power density is controlled through a 'design rule' limit that is analogous to the electromigration-driven current limits, with the exception that it is based on the rms rather than average current value. With high performance, deep sub-micron products, this type of current limit is a real limitation in design and needs to be managed carefully.

- Temperature is controlled by the power dissipation of the chip, the package, and by the system cooling environment. Here, too, there are cost and performance trade-offs that need to be managed.

8. ALPHA PARTICLE SENSITIVITY

Absorption of ionizing radiation has been shown to cause a disturbance in the operation of semiconductor devices [Matsumoto 1981]. For example, a single alpha particle penetrating the surface of a semiconductor device can generate up to 1 million electron-hole pairs, a charge surge that is sufficient to change the state of a storage node such as a RAM cell. This disturbance, known as a "soft error," is a random, non-recurring, single event and is different from the permanent "hard" damage caused by other failure mechanisms, including radiation-induced latchup or threshold drift.

Ionizing radiation, such as alpha particles, cosmic rays and some types of X-rays, can all cause soft errors. Traditionally, [May 1978] the principal cause of the soft errors is alpha particles emitted from trace levels (parts-per-million) of radioactive isotopes of uranium and thorium present in many materials used in microelectronics packaging.

8.1 Description of the Mechanism

As indicated, soft errors are caused by the charge generated in silicon by the ionizing radiation. That is, for example, an alpha particle penetrates into silicon and generates a cloud of electron-hole pairs. These carriers then diffuse through silicon, and either recombine or are collected by a p-n junction (if they are generated within a diffusion length from the junction). The carriers that are collected are separated by the junction electronic field, and cause a current pulse. This pulse of current may cause a node to toggle. An obvious case is a RAM storage node, programmed to store a 'zero', that collects sufficient charge generated by the ionizing radiation to appear to be storing a 'one', i.e., causing an error. The error is, however, erased on the next programming or refresh cycle, and is therefore 'soft.' The ensuing Soft Error Rate (SER), caused by this bombardment of silicon by the ionizing radiation, is a key metric of circuit stability and reliability.

There are three major conditions that alpha particles must satisfy in order to cause a "soft error":

1. Alpha-emitting radioactive isotopes must be present near the interior of the package material to allow escape of high energy (MeVs) alpha particles before they are internally absorbed. Typically these isotopes must be within 50 mm of the device surface. This condition really specifies that the probability of interaction, i.e., the soft error rate, is directly related to the raw number of alpha particles present.

2. Emitted alpha particles must have a direct line of sight to the active device, i.e., the probability of interaction is directly related to the number of alpha particles absorbed in the vicinity of a target device.

3. The target device must be sensitive to current pulses generated by the alpha particles.

The soft error rate is then a function of the alpha particle flux, the solid angle described by the geometric relation between the ionization source and visible silicon area, and the sensitivity of the devices in the visible area.

8.2 Modeling of the Mechanism

Alpha particle flux is defined as the number of alpha emissions from the surface of a material per hour per square centimeter of area. The emmissivity of a material, defined as the alpha particle disintegration rate per unit weight, is usually reported in units of picoCuries/gm. The most common method of reporting alpha particle content in the semiconductor literature is as alpha flux rate. As a "rule of thumb", an alpha flux of one alpha/cm^2/hr corresponds to approximately one ppm of uranium in a material. It has been reported that even flux levels below 0.1 alpha/cm^2/hr can give rise to significant soft error rates in semiconductor memory devices [May and Murray 1978]. Trace amounts of uranium and thorium in many of the materials used in the manufacture of microelectronic packages [May and Murray 1978] are a major cause of alpha particles, and are the principal cause of the soft errors in semiconductor memory devices.

In general, semiconductor devices are contained in one of the two types of packages: ceramic or plastic. Ceramic packages are typically hermetic packages for high performance, high pin-count and high reliability applications. Plastic packages are non-hermetic packages for low cost and high volume applications.

Hermetic ceramic packages consist of an alumina package body with a cavity containing the die, and a metal or ceramic lid attached to the alumina body by a metal or glass seal. Each of these materials contains radioactive contaminants. Commercial alumina substrate (flux < 0.1 alpha/cm^2/hr) and the package lid (flux ~ 0.1 alpha/cm^2/hr for gold plated kovar or alloy 42 or several alpha/cm^2/hr for ceramic) are the most critical materials because they lie in a plane parallel to the die surface. The glass used to seal the hermetic package often contains zircon at concentrations up to 20% which may contain 100 to 1000 ppm of uranium. However, the low exposure angle and the small width of exposure of the sealing material limits its contribution to alpha particle emissions to less than 0.1 alpha/cm^2/hr.

In transfer molded plastic packages, only the polymeric material and the filler come into contact with the die. The polymeric material, an organic hydrocarbon, is relatively free of radioactive materials and does not contribute to the alpha particle emissions. The filler materials include fused silica, crystalline silica and alumina at concentrations of 70 to 80% by weight. Standard silica and alumina fillers typically contain several ppm of uranium and thorium impurities, but natural quartz sources discovered in India and Brazil can be used to produce low alpha fused silica with 10 to 50 ppb of uranium, corresponding to an alpha flux of only 0.01 to 0.05 alpha/cm^2/hr. Synthetic fused silica filler is available with 0.5 ppb levels of uranium and thorium and alpha fluxes below the detection limit of 0.001

alpha/cm^2/hr. It is hence a relatively simple task to reduce alpha emission levels in transfer molded plastic packages by the use of synthetic fillers.

The continued scaling of dimensions and increasing integration levels mean that the SER sensitivity of IC devices is increased. That is, the critical charge of a RAM cell is reduced with normal dimension scaling, and hence smaller ionization current pulses may cause soft errors. This makes modern circuits sensitive not only to lower doses of alpha radiation, but also to cosmic rays. Since cosmic rays cannot be managed through material control and cannot be easily blocked, a maximum desirable soft error rate dictates the magnitude of minimum critical charge, and hence dictates the minimum size of a RAM storage cell.

8.3 Prevention of Alpha Particle-Induced Damage

The soft error rate can be managed by either reducing circuit sensitivity to current pulses, or reducing the critical flux of ionizing radiation. There are several ways of accomplishing this, and some or all of these are routinely used:

- Reduction of circuit sensitivity through chip design. There are a number of design tricks, such as error detection and correction through parity checks, and careful circuit layout, that can be employed at RAM architecture or circuit design level to reduce SER.

- Reduction of circuit sensitivity through chip process. There are a number of process tricks that can be used to increase the critical charge stored in a node, and hence to reduce the sensitivity to current pulses. These range from crafting high-C storage cells through appropriate tailoring of the diffusion profiles to using trenches in silicon to create large vertical capacitors with small footprints. In addition, a grid of dummy junctions can be created below the active area to sweep the generated charge away from the critical circuits.

- Reduction of ionizing flux through material control, i.e., elimination of radioactive contaminants from the packaging materials. In the case of ceramic packages this may be too costly. As discussed above, for molded plastic packages this is quite effective.

- Reduction of ionizing flux through packaging process. It is possible to shield IC devices from ionizing radiation by coating the die with polyimide films. This approach, however, may lead to other reliability problems, due to poor adhesion or stresses from the polyimide coating, and may not be compatible with larger die sizes required for MBit RAMs.

- Reduction of ionizing flux through package design. It is possible to minimize the impact of ionizing radiation by tailoring the geometry of the package so as to reduce the angles between the radiation sources and sensitive die areas.

9. ELECTROSTATIC DISCHARGE AND ELECTRICAL OVERSTRESS

Electrostatic discharge (ESD) is defined as the transfer of charge between two bodies at different electrostatic potentials. This charge transfer results in a pulse of energy that can cause several types of electrical overstress (EOS) failure modes. The EOS failure modes seen in IC components include shorting due to damage of the dielectric layers or device junctions, or opens due to damage of the metallization runs. Dielectric layers and device junctions are damaged by high voltage (low current) overstress. Metallization damage results from high current (lower voltage) overstress, due to resistive heating of a structure (up to and above melting temperatures) that is so rapid that the energy cannot be dissipated before damage occurs.

Electrostatic charge is generated by either frictional contact between dissimilar insulators, known as triboelectric charge generation, or by the transfer of charge between a conductor and an insulator, known as induction charging.

EOS is typically defined as an event arising from inappropriate use of a component, typically during IC or system test procedures. If, for example, the test protocol forces a component pin to some potential (e.g., ground), while the logic state of the circuit drives that pin to some other potential (e.g., Vdd), a contention state is established that results in a Vdd-to-ground short across that pin. This condition, often transient, results in large current surges that can create significant damage. Similar contention states can be established within the circuit itself, if it is not specifically addressed in the chip design. EOS events are typically slower than ESD events, and involve a transfer of larger amounts of energy.

Pareto ranking of the failures in electronic components shows that ESD and EOS are major contributors. ESD/EOS are not only quality issues, affecting the components during manufacturing, testing and/or shipment, but are also principal reliability concerns because they often result in partial damage of the components. The partially damaged circuits then fail while in use, typically in the infant mortality portion of the product life cycle, due to interaction with other failure mechanisms. Addressing these problems affects many aspects of the components including their design, shipping and handling, and production and testing. The significance of ESD and EOS damage increases with the scaling of the chip feature dimensions, because the circuit sensitivity to these phenomena typically decreases. In addition the immunity of the MOS devices to these events is lower, and their increased usage has thus also increased the frequency of ESD/EOS failures.

9.1 Description of the Mechanisms

Electrostatic discharge damage is a specialized case of excessive voltage damage in which the overall energy of the transient is small despite extremely high voltages. The ESD pulse can be a discharge of several 1000s of V over a few ns periods. This type of a pulse is an 'arcing' phenomenon that can easily damage dielectrics or junctions in an IC. The damage typically is quite localized, appears like a pin-hole in the material, and results in a short across an insulator, or a junction. Whereas the damage may be difficult to detect through physical failure analyses, ESD 'signature' is that the effect is in the periphery of the circuits (normally in the I/O buffers), and, in case of handling induced ESD, affects circuits connected to the pins that are most exposed to casual contact, typically at the extremes of the package outline.

Electrical overstress damage is typically much more extensive than that associated with ESD. Extensive metallization line damage and melting, blown wire bonds, and other destruction, often referred to as a 'smoking hole', are not uncommon. As indicated, EOS damage can be induced either during a test where the current is provided by the test equipment, or in use, where the current is provided by the system power supplies. With the very large circuits consisting of millions of logic gates, made possible by modern deep sub-micron technologies, both types of ESD/EOS are becoming increasingly probable. For example, the conditioning of such components, required to put the circuit in some known logic state, may call for precise sequencing of 100s or 1000s of test stimuli to be delivered through 100s of pins, a situation where a contention state may easily be established. Similarly, it is often not possible to estimate the timing on all the logic paths within such a circuit, so that establishment of a contention state within a circuit is increasingly probable, too.

9.2 Modeling of the Mechanism

There are three models to describe different ESD phenomena that occur in electronic components:

- The human body model (HBM) that simulates the effects of human interaction with the components. The HBM is represented by a discharge through a series RC circuit with the resistance of 1500 Ω, and the capacitance of 100 pF. A person can, through the triboelectric charge effect build up 1000s of V by simply walking on an insulated surface (e.g., carpet), or using insulated footwear, given the right set of environmental conditions (typically low humidity). The characteristic waveform of the discharge associated with a person coming in electrical contact with a body at some other potential is then modeled by the HBM circuit.

- The machine model (MM) that describes the effects of charge transference between machines and components. The machine models are intended to model the ESD produced when contact is made between a charged object and an IC during the manufacturing process, at steps such as wire bonding, board assembly, or test. Unlike the human body model, industry does not have a long tradition with the machine model specifications, tests, and cures. Therefore, two specifications originating from Japan and Phillips Research Laboratories are used

Generally MM is simulated by a discharge through a RC circuit with the resistance of 0 Ω, and the capacitance of 200 pF.

- The charged-device model (CDM) that describes the effects of rapid discharge of the charge collected in the device itself. An electronic package can pick up a charge from the triboelectric interaction with, for example, the shipping tubes or marking equipment. When the package is subsequently grounded this collected charge dissipates to ground. Whereas the CDM has not been fully standardized throughout the industry, it is generally simulated by a discharge through a RC circuit with the resistance of 0 Ω, and the capacitance of 0 pF.

A set of test procedures is normally used to evaluate the ESD 'immunity' of a component, i.e., to evaluate the magnitude of an ESD discharge that a component can absorb without permanent damage. These procedures normally call for delivering successive simulated ESD zaps to each component pin, and increasing the discharge voltage (current), and alternating the polarity, until damage is detected (usually by measuring pin leakage). Specialized equipment that performs this type of testing is commercially available from a number of vendors.

9.3 Avoiding ESD/ EOS Failures

There are two basic ways to manage the ESD damage, either through elimination of the ESD events or through enhancement of the device ESD immunity.

- Prevention of the ESD events calls for establishing appropriate manufacturing environment, including installation of humidity control.

- Systems and conducting floors, establishment of handling procedures and tools (e.g., use of conducting footwear and grounding wrist straps), and institution of equipment maintenance protocols. These disciplines are all required, and are fairly endemic throughout the industry. Clearly, the cost of *not* minimizing the ESD events exceeds the obviously high cost of implementing these systems and practices. This preventive approach is never perfect, and must be used in concert with enhancement of device immunity.

- Increasing device ESD immunity is typically accomplished by designing protective circuits into the ICs. These chip-protection circuits are typically intended to provide alternate paths that safely divert ESD charge to ground. The design of the ESD protection networks can often be almost an art form, because their response must be rapid enough to react to the fast ESD pulses, but they must not affect the normal desired operation of the circuit. In addition, the volume of the ESD protection network must be sufficiently large to absorb and dissipate the zap energy without incurring any damage, but the area of the circuit needs to be compatible with chip size and cost targets. The design of the ESD protection circuits is often augmented by implementation of a separate set of layout design rules that dictate the use of different feature dimensions in the I/O buffers rather than in the rest of the component. The continued scaling of dimensions into deep sub-micron regions makes the implementation of these circuits incrementally more difficult. Many modern silicon processes include separate steps to facilitate the implementation of the ESD protection circuits.

There are no specific techniques for preventing EOS other than practicing careful engineering through component design and test cycles.

10. LATCHUP

Latchup is defined as a situation where a sustained low impedance parasitic path is created between power and ground rails. If the low-impedance path persists for long enough, latchup is a destructive phenomenon that often results in blown bond wires and massive widespread destruction of the chip. Latchup is caused when a parasitic SCR, associated with a pair of either bipolar or MOS devices, giving an arrangement of four alternating diffused areas (p-n-p-n), is triggered. This phenomenon is impacted by a number of process, layout, and circuit factors including layer sheet resistivities, diode voltages, leakage and impact ionization currents, device gains, transient wave forms and Vdd sag.

Dimension scaling, higher integration levels and higher operating frequencies, all tend to aggravate latchup, so that aggressive control measures are required to avoid latchup in deep sub-micron products.

10.1 Description of the Mechanism

P-N-P-N devices are susceptible to latchup if they form a n-p-n and a p-n-p loop that are closed, and have a gain that is larger than 1. Then, any trigger that turns on one of the two devices will result in a latched state. If the resistances associated with the latched circuit are sufficiently low, the voltage (current) required to sustain the latch will be below the applied Vdd value, and the circuit will self destruct.

As shown in Figure 10.1, a standard CMOS gate consisting of an n-MOS and a p-MOS can form a latchup pair. The sketch shows a case with a p-well and an n-substrate, but a similar situation arises in n-well devices using a p-substrate. Typically, the well is relatively shallow (few microns) due to several process and circuit performance reasons, so that the gain of the vertical n-p-n is of the order of 10s. Space considerations typically dictate that the distance from the p+ to the n+ is of the order of a micron, too, giving the lateral transistor a gain that is somewhat below 1. But the product of the gains is normally greater than 1, so that the CMOS configuration lends itself to latchup. R1 and R2 represent lumped values of the total distributed resistances in the well and substrate, as seen by the latch circuit. Thus, with resistive regions (high R1 and R2), small amounts of leakage current may cause a sufficient IR drop to forward bias an emitter-base junction and thus trigger latchup. On the other hand, with conductive regions (low R1 and R2), the holding voltage will be below Vdd and the latch will be sustained until the circuit self destructs.

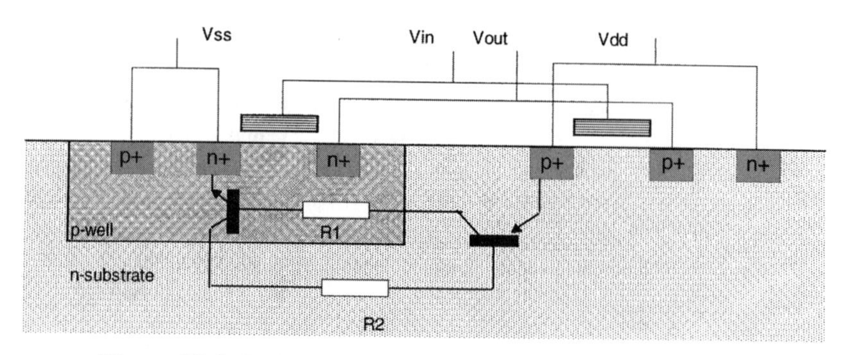

Figure 10-1 CMOS transistor pair, wired up as a primitive inverter

In Figure 10-1, the well is shown explicitly connected to ground, while the substrate is connected to Vdd, which is usual for digital circuits. With analog circuits, a separate power supply is generally used to provide equal positive and negative voltage swings around ground. In addition, with analog circuits the n-MOS device may be biased into the region that maximizes the generation of the well (substrate) current caused by impact ionization in the drain electronic field. This can clearly further complicate managing latchup in analog devices.

Latchup is a reliability issue, as well as a time-zero concern, because:

- The IR drops across the well and/or substrate may drift with time, eventually turning on a transistor, either through increasing leakage current caused, for example, by spiked junction, or degrading plug contact resistance, caused, for example by electromigration; or

- A random phenomenon, such as voltage spike or sag, noise, temperature spike, current spike caused by some ionizing radiation, and uncontrolled power-up sequencing, may trigger latchup.

In either case, i.e., whether latchup is caused by some genuine drift mechanism or by a random combination of external effects, it 'feels' like a reliability problem to a user, and must therefore be addressed. Most often, latchup is triggered by some external effect, but the sensitivity to the latch may degrade over time through some intrinsic reliability mechanism. Therefore, latchup failures may be expected to occur anywhere in the bathtub curve.

10.2 How to Detect

As indicated above, latchup is often triggered by a random phenomenon, so that it is difficult to measure the 'probability' of a latchup event. It is, however, possible to assess, in a qualitative way the component immunity to latchup through standard tests and/or procedures.

- I/O Buffers. Traditionally, chip I/O buffers, being the interface to the outside world, and hence being the most exposed to random phenomena that trigger latchup, have been considered to be the circuit elements with the highest latchup risk. A standard test, where a known current (typically 100 mA) is injected into a buffer, by forward biasing a junction, has been evolved and is routinely used.

- Chip Core. There are no standard tests for evaluating the margins of a component to a latchup event somewhere in the core of the chip. Good product engineering practices, however, typically define final test conditions which do assess latchup margins; e.g., tests at elevated or reduced voltage, high temperature tests, and skew tests. Burn-in can also be a good method for assessing latchup margins.

10.3 How to Avoid

There are several fundamental strategies for managing latchup.

- Minimize the probability of triggering latchup. This is accomplished mostly through implementing a 'clean' design throughout the system, so as to minimize all sources of spikes that may trigger a latch. Typically, this method calls for careful design of the power and ground distribution network at chip, package, board and system level, as well as management of signal switching density and rate. Clearly, whereas these goals constitute good engineering practices in general, it is not possible to guarantee absence of all triggers under all circumstances.

- Maximize circuit immunity to latchup triggers. There are a number of design and process tricks that are commonly used to achieve this. Use of single or even double 'guard rings,' diffusions that separate n-MOS and p-MOS devices, is the most common design practice. This approach, however, is costly in terms of chip area and is hence typically reserved only for nodes at highest risk of latchup, such as I/O or analog circuits, or for IC applications requiring latchup immunity during large surges of radiation. Use of well and/or substrate 'taps,' ohmic contacts connected to the Vss/Vdd distribution network, to minimize the distributed resistivity of the well and/or substrate regions, is also a common practice. In fact, the frequency of taps, as well the limitations on specific layout configurations, is often specified as one of the design rules. The primary process technique for managing latchup is the use of highly doped substrates to provide a good shunt to ground, below an epi layer needed to provide desired background doping for active devices. Use of trenches or even full dielectric isolation between the devices in order to improve their decoupling, or use of various doping techniques to reduce the resistivity of the well/substrate regions is also relatively common.

- Maximize the holding voltage, i.e., minimize the circuit's ability to sustain a latch. As indicated, selection of appropriate R1 and R2 values can force the holding voltage to be above Vdd. Use of appropriate separation between p-MOS and n-MOS devices, often dictated by the design rules, can also ensure this.

11. DESIGN FOR RELIABILITY

As indicated throughout this book, IC component failures are affected by the design, the manufacturing process, the use conditions, and the interaction between the three. The physics of failure methodology can be used to model the effect of these factors and provide guidance to control them for maximum reliability. The physics of failure (PoF) approach to design-for-reliability is founded on the fact that the failure of electronics is governed by fundamental mechanical, electrical, thermal, and chemical processes. By understanding the possible failure mechanisms, design teams can identify and solve potential reliability problems before they arise.

The PoF approach begins with the first stages of design, as is shown in Figure 11-1. A design team defines the product requirements, based on the customer's needs and the supplier's capabilities. These requirements can include the product's functional, physical, testability, maintainability, safety, and serviceability characteristics. At the same time, the application environment is identified, first broadly as aerospace, automotive, business office, storage, or the like, and then more specifically as a series of defined temperature, humidity, vibration, shock, or other conditions. The conditions are either measured, or specified by the customer. From this information, the design team, usually with the aid of a computer, can model the mechanical, electrical, thermal and chemical stresses acting on the product.

Next, stress analysis is combined with knowledge about the stress response of the chosen materials and structures to identify where failure might occur (sites), what form it might take (modes), and how it might take place (mechanisms). For semiconductor devices, failure mechanisms include electromigration, time-dependent dielectric breakdown, stress-driven diffusive voiding, and hot-carrier degradation. Once the potential failure mechanisms have been identified, mathematical models for these mechanisms are used to evaluate the time-to-failure for the component in any environment as a function of the operating stresses on the part, the environmental stresses on the part, and the part architecture, including the geometry and the materials of construction. This evaluation identifies the dominant failure mechanisms (i.e., those predicting the least time-to-failure) and provides information on whether the component will survive its intended application life. The design team then evaluates the sensitivity of the time-to-failure by the dominant failure mechanism to changes in the design. In this way, design tradeoffs which maximize reliability and increase robustness can be determined either iteratively or with the use of optimization software.

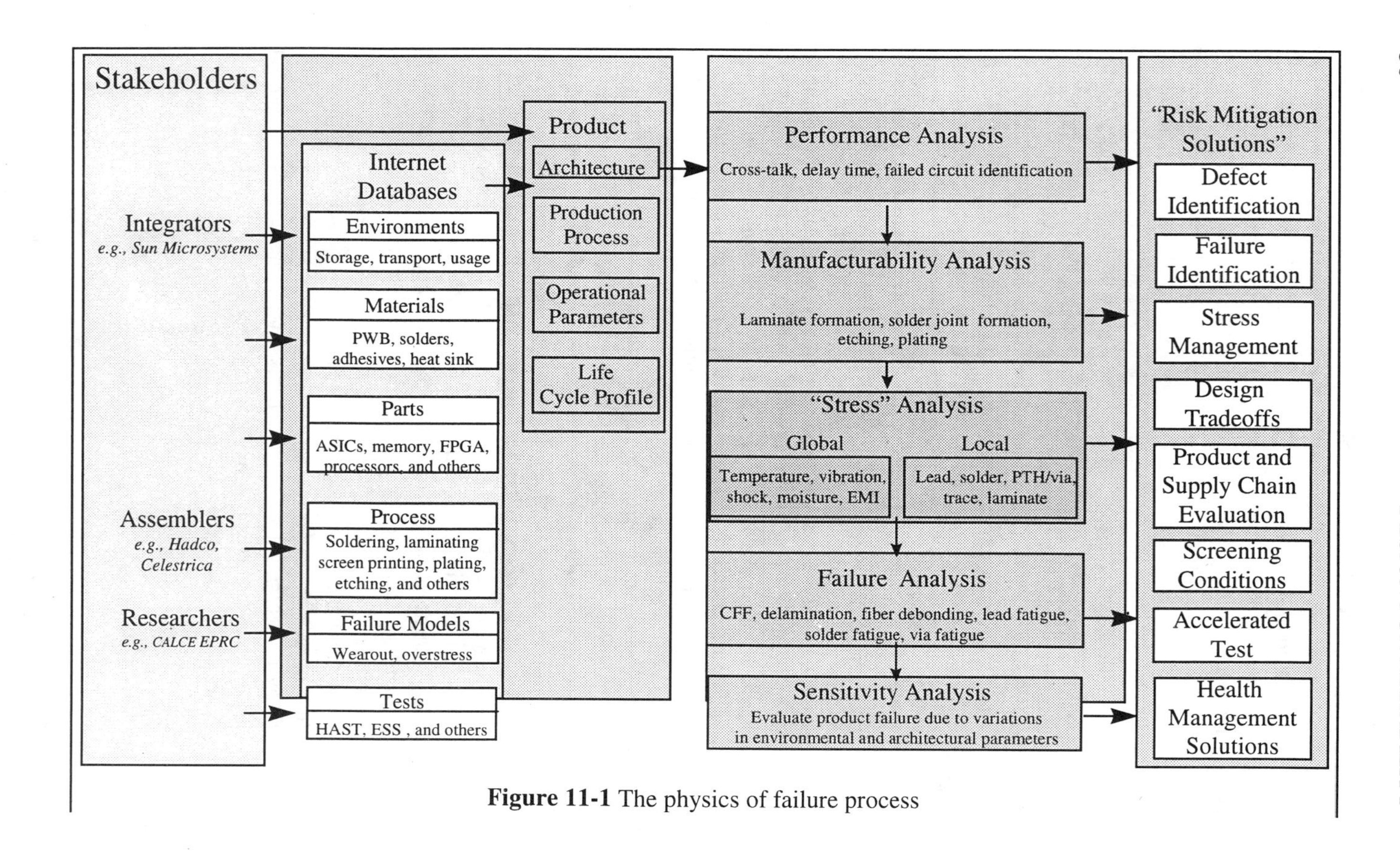

Figure 11-1 The physics of failure process

In addition, values for the operational and environmental stresses can be selected for qualification testing which will accelerate failure of the device without altering the failure mechanism. The physics-of-failure approach is also used to qualify manufacturing processes to ensure that the nominal design and manufacturing specifications meet or exceed reliability targets.

The physics-of-failure process can be extremely complex and so requires the use of an expert system for its completion. Computer software has been developed by the CALCE EPRC at the University of Maryland to make physics-of-failure based design, qualification, planning, and reliability assessment manageable and timely. The details of this method and the characteristics of the expert systems to facilitate its use are described in the following sections.

The reliability of an IC component over time can be described as a function of two phenomena, both of which can be evaluated using physics-of-failure techniques. These are 1) wearout failures, which are characterized by an increasing failure rate, and which are caused by material wear, fatigue, and other intrinsic aging mechanisms; and 2) infant mortality, which is characterized by a decreasing failure rate, and which is caused by manufacturing defects and/or specific design sensitivities. The expert system for design-for-reliability, which is described in the following sections, addresses both of these phenomena.

11.1 Design System

In order to create modern, deep sub-micron components, a structured design system is required and can be used to implement a practical design-for-reliability methodology, such as is described here in Sections 11.1-11.5, after Oshiro et al. [1992]. This methodology relies partially on use of structured Design Systems, but is mostly an implementation of a system of practices based on PoF. Such systems for design are often hierarchical and divided into three distinct parts, as illustrated in Figure 11.2. During the 'development phase,' a cell library database is generated which includes all functional, timing and test models, layouts and schematic databases, etc., along with an integrated CAD flow, and new CAD tools when necessary. During the 'chip design phase,' a floorplan, clock tree and power distribution network, etc., are defined and a specific IC is designed by implementing cells from the library. Test pattern generation, circuit simulation and logic verification are performed during the 'test development phase,' and are usually executed in parallel with chip design.

Design-for-reliability (DFR) tools and procedures can be used to supplement all three phases, leveraging the data and procedures already available in the design system. This results in a qualified cell library, qualified chip design rules and a set of tailored qualification tests that ensure part reliability. Thus, chip design reliability is implemented using the procedures described here, and summarized in Figures 11.3, 11.4 and 11.5.

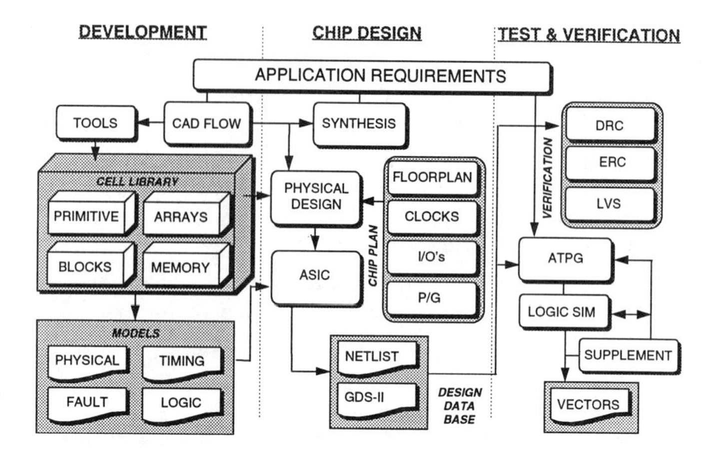

Figure 11-2 A hierarchical design system showing the general activities in the three phases of the design flow.

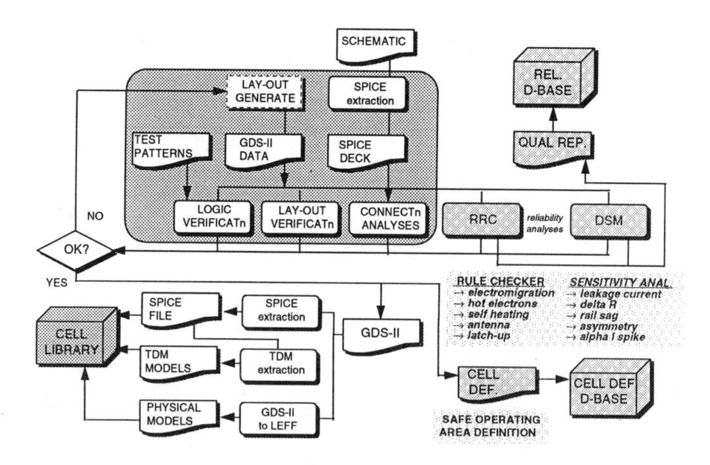

Figure 11-3 DFR activities during the library development phase illustrating the required input data and the available output data.

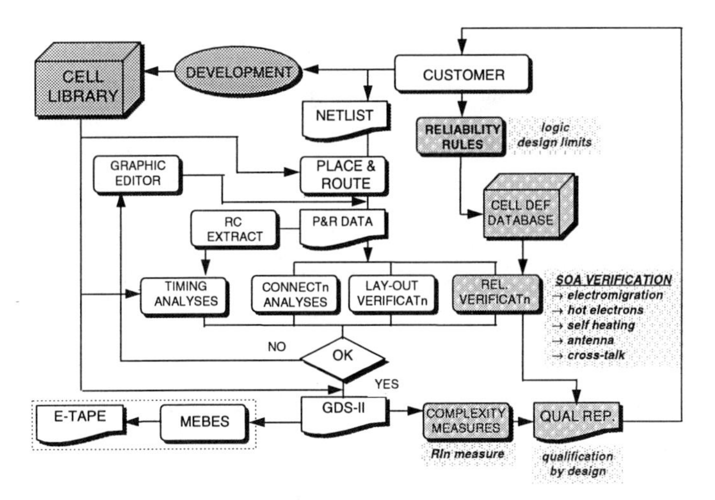

Figure 11-4 DFR activities during the chip development phase illustrating the required input data and the available output data.

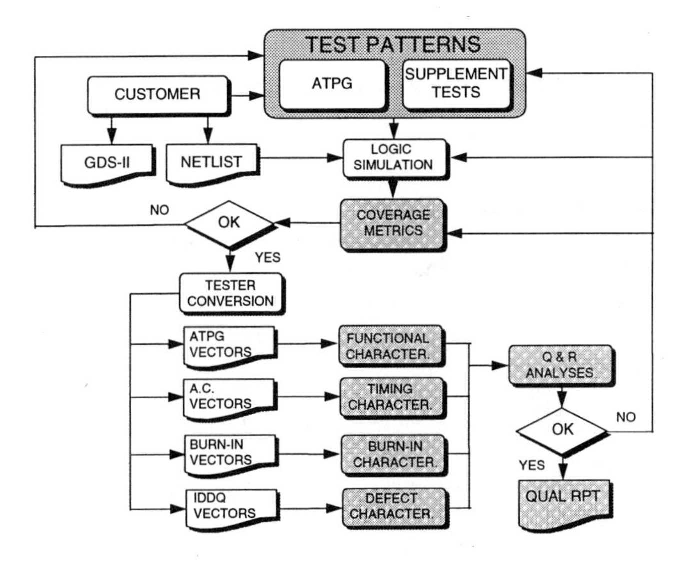

Figure 11-5 DFR activities during the test development and verification phase illustrating the required input data and the available output data.

11.2 Effective Management of Wearout Failures

Many currently available software programs try to address reliability using outdated and inherently flawed techniques such as MIL-HDBK-217 [Cushing et al. 1993, Watson 1992, Bar-Cohen 1988]. These systems are doomed to produce incorrect results and ineffective design solutions. Designs which effectively manage failures must be constructed using the principles of physics-of-failure (PoF). The software which facilitates the use of PoF techniques may manage failures by conducting on-line sensitivity analyses of the effects of design on failure of the part by the dominant mechanisms or through the use of "reliability rules" formulated on the basis of physics-of-failure analysis.

Some of the wearout failure mechanisms which can be addressed in this way include:

- Electromigration: Metallization line widths and thicknesses, operating temperature, and current density can be adjusted to improve reliability without sacrificing performance.

- Time-dependent dielectric breakdown: Oxide thicknesses, gate voltages, and operating temperatures can be adjusted along with manufacturing parameters for oxide growth and/or deposition.

- Stress-driven diffusive voiding. Passivation deposition temperature, metallization width and thickness, and operating temperature can be adjusted.

An alternate, but also effective, route of addressing these failures is through a set of Reliability Design Rules based on physics-of-failure that ensure wearout occurs beyond the useful life of a system. In this case, conventional practices for characterization of failure mechanisms, utilizing accelerated tests on discrete test structures, are used to generate the 'Reliability Rules' that define the capabilities of a given process technology. Then, compliance with these rules for every node on every design is verified hierarchically via 'Reliability Rule Checker' (RRC) tools and procedures. Sample failure mechanisms which can be addressed in this way are the following: intrinsic electromigration, stress voiding, hot-electron aging, antenna induced TDDB and latchup. The methodology used is described in Figure 11.6 and the associated text.

- Current Density Mechanisms: Electromigration and Self-Heating. The RRC for the current-driven mechanisms relies on the hierarchical nature of the designs. Verification is performed at the cell level first by checking structures entirely within the cells at the schematic level.

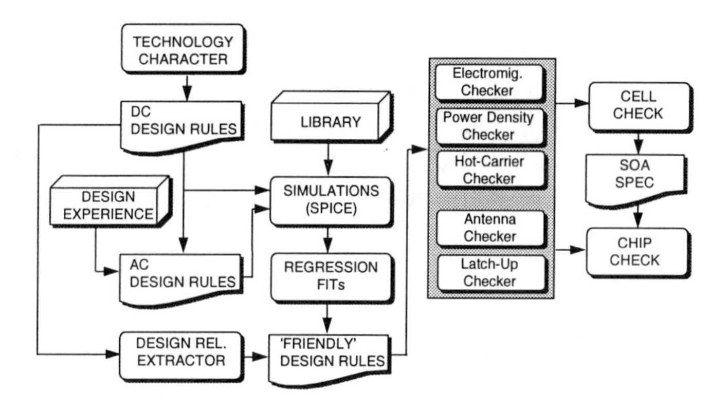

Figure 11-6 The development and implementation of the 'Reliability Rule Checker' tools, starting from the basic process technology assessment through to automated tools.

- A gross design 'envelope' is defined, to allow rapid filtering of the internal nodes that easily meet all the rules. Nodes outside the envelope are analyzed in more detail by calculating a "worst case" current based on various circuit parameters (e.g., device sizes, slew rates, capacitive load) and determining the design rule limit based on the layout. The interconnect between the cells and the I/O portions of the cells are checked using an equation which is based on a regression fit to SPICE results. These equations describe the 'Safe Operating Area' (SOA), where the reliability rules are met for a given cell, in terms of environmental (temperature, frequency) and layout (load capacitance, layout feature) factors. Compliance with these global limiting factors is then the only thing that is checked at the chip level.

- Hot-Carrier Aging: Hot-Carrier Aging design rules are typically defined in terms of substrate (I_{bb}) and drain (I_{dd}) currents. The methodology that is followed is analogous to the electromigration rule checks. The design rules are dependent on circuit and layout, and are checked hierarchically.

- In-Process Electrostatic Discharge: The Antenna Effect is a factor that can significantly impact TDDB failure characteristics. This plasma-induced damage is often managed through a set of layout-based design rules that limit the ratio of the conductor (antenna) area and the gate area. An RRC tool based on a layout analyzer that extracts physical design information (Dracula) and a post-processor that determines the required area ratios have been developed. The post-processor also checks the conductors for a diffusion connection which nullifies the 'antenna' effect. Since the antenna can typically be on any one of several metal levels, this layout analyzer extracts connectivity information at each of these levels. The antenna check is run at both the cell and chip levels.

- Latchup: A multi-parameter phenomenon impacted by a number of process and layout factors that control sheet resistivities, diode voltages, parasitic bipolar gains, etc. However, latchup is often controlled by managing IR drops, such that the controlling *layout* factor is the frequency of the grounding taps. Thus, the design rule is often defined in terms of the well-tap to gate distances. An RRC tool based on a layout analyzer and a post-processor, as in the antenna checker, is used to determine and check the well-tap to gate distances and to identify any violations to the rule. In a hierarchical design system, the latchup check is performed only at the cell level since no additional active devices are defined at the global level.

11.3 Extrinsic Reliability Mechanisms

Extrinsic failures are an attribute of both the defectivity associated with the silicon process and the complexity of the design and real world stresses. It is assumed that the chip failure rate, the time constant bottom of the bath-tub curve, is a statistical representation of the interaction between these three attributes. The design attribute that impacts the extrinsic failure rate is then purely an expression that describes the complexity of the total design and manufacture process.

This is managed through the use of "Reliability Index Numbers" (RINs), a quantitative description of all topographic features that statistically interact with reliability (e.g., number of contacts or vias, total active area, total length of minimum width metals, etc.). For a given process technology, a list of RINs is defined through a combination of historical experience, in-process defect analyses and product yield analyses. An example of a RINs list is shown in Table 11.1. A layout analysis used to quantify each RINs for a given chip and a linear scaling based on RINs counts is performed to extrapolate a given product failure rate from measured failure rate based on test vehicle life tests. The failure rate may be adjusted to reflect maturity of a process or an IC fabrication facility.

Table 11-1 A Sample Listing of Selected Defects and Tolerances, Showing their Relationship to the Root Cause Failure Mechanism and Failure Mode

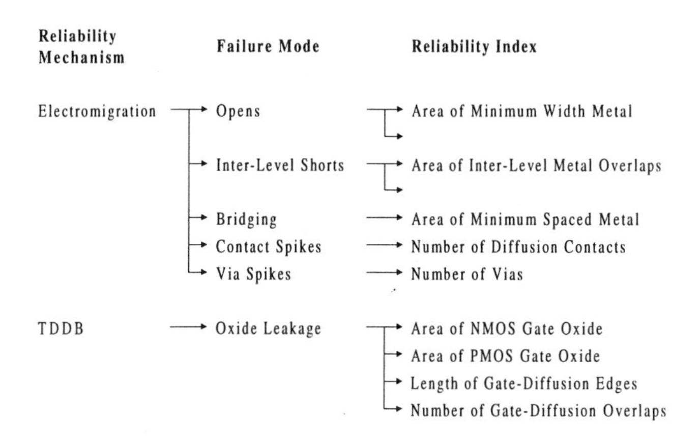

Reliability Mechanism	Failure Mode	Reliability Index
Electromigration	Opens	Area of Minimum Width Metal
	Inter-Level Shorts	Area of Inter-Level Metal Overlaps
	Bridging	Area of Minimum Spaced Metal
	Contact Spikes	Number of Diffusion Contacts
	Via Spikes	Number of Vias
TDDB	Oxide Leakage	Area of NMOS Gate Oxide
		Area of PMOS Gate Oxide
		Length of Gate-Diffusion Edges
		Number of Gate-Diffusion Overlaps

11.4 Infant Mortality Failure Mechanisms

Failure by infant mortality is related to both the number and severity of the defects associated with the silicon process and the failure mechanisms which are associated with these defects. For a given process technology, a list of the potential defects is defined, through a combination of historical experience, in-process defect analyses and product yield analyses. An example of such a list is shown in Table 11.1. Layout analyzers are used to quantify the number of possible locations at which a defect may occur for a given chip and a calibration of the failure mechanism model is performed using actual test data to determine the probabilistic distribution of defects as defined by the number of defects of each size.

The failure mechanism models can be used to determine the effects of changes in the probabilistic distributions of defects on the reliability of the component and thereby be used as a way of determining the optimal defect levels which preserve reliability at a reasonable manufacturing cost. This information combined with a knowledge of the effect of manufacturing process changes on defect densities can be used to conduct robust processes and to select optimum process parameters.

11.5 Circuit Sensitivities

It is important not only to understand the failure of components due to the catastrophic failure of a device element such as a metallization short or open or a breakdown of the oxide, but also due to the accumulation of a number of parameter shifts, none of which would be sufficient to cause failure independently. This requires that sensitivity analysis be conducted at the integrated circuit level and that 'Design Sensitivity Metrics' (DSMs) be used to manage this issue in a manner analogous to the Reliability Rule Checks. Whereas the RRCs described in Section 11.1 are based on known failure mechanisms with defined 'pass/fail' criteria, the metrics described here are based on comparison to the established and evolving experience base. This is accomplished through a multi-step process that is continually evolving, as illustrated in Figure 11.7. An envelope of DSM metrics is defined based on general reliability experiences, and is then honed and redefined through a series of comparisons with the real products to represent a more realistic set of criteria. In addition, new DSM metrics are defined with every new test, system or field experience.

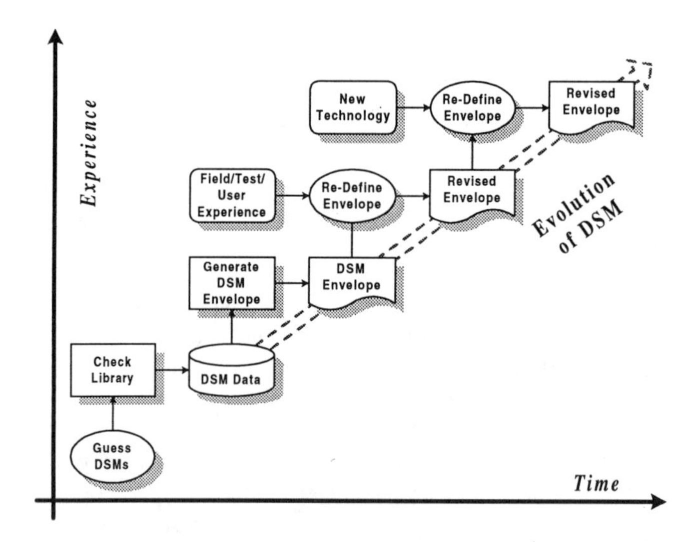

Figure 11-7 Evolution of the "Design Sensitivity Monitor" (DSM) metrics, defined to assess the robustness of designs and to relate this to the in-field experience base.

- Leakage Sensitivity: Sensitivity to defect-induced leakages is characterized by determining the amount of leakage current required to cause a functional failure in the circuit. This is accomplished by inserting a variable resistive path in the circuit at the sensitive node(s). It has been determined that dynamic circuit nodes, or nodes driven by weak devices, are most susceptible to leakage.

- Metal Conductance Degradation: The degradation of conductance of the metal interconnects (caused by electromigration or stress voiding) is simulated by inserting a resistor in series between a driver and the capacitive load of a given cell. This typically results in performance degradation. The amount of degradation for a fixed R is used to characterize the circuit sensitivity.

- Transconductance Degradation: While hot-electron aging rules control the amount of aging per node, the sensitivity of a node to the resulting transconductance degradation is treated as a DSM. This is done by decreasing the device width for the aged device and rerunning the simulation. The design sensitivity is characterized by the degree of performance degradation.

- Temperature and Voltage Effects: Since the performance of a CMOS circuit is affected by operating temperature and/or voltage, design verification simulations are run at nominal temperature and voltage as well as at degraded (higher) temperature and (lower) voltage. The design sensitivity is characterized by the degree of performance degradation.

- Alpha Particle Sensitivity: Immunity to alpha particle strikes is evaluated by simulating sensitive circuit elements (memory cells, latches) with a piece-wise linear current source. The design sensitivity is characterized by the amplitude of the current required to flip the state of the cell.

- Noise Sensitivity – Slew Rate Characterization: In digital circuits, signals with slow slew rates are the most susceptible to 'noise' defined here as the cross-coupling of an active signal onto a passive signal during a transition. Thus, this sensitivity is characterized by the node slew rate.

- Noise Sensitivity – Voltage Margins: Circuits with dc current flow, such as drivers with passive devices, or other applications where a node is not driven purely rail-to-rail, are susceptible to noise perturbations, due to their sensitivity to voltage margins. This sensitivity is measured by simulating the node with an added voltage source in series.

12. PROCESS DEVELOPMENT

12.1 Introduction

Multiple levels of metallization are needed when, for technical and economic reasons, interconnects cannot be laid out efficiently in a single level and therefore need to be stacked one on top of the other [Davis 1989, Martinez-Duart and Albella 1989, Geraghty and Harshbarger 1990, Wilson et al. 1988, Mitsuhashi et al. 1989, Small and Pearson 1990, Bartelink and Chiu 1989, Osinki et al. 1989, Oakley 1985, Pai and Ting 1988, Saxena 1988, Kotani 1988, De Santi 1990, Sum et al. 1987, Cagnoni et al. 1989, Fisher et al. 1989, Nishida et al. 1989, Saxena and Pramanik 1984, Wolf 1990, Nishida 1989]. The technical issues in interconnect layout stem from increased density of transistors that need to be connected, while economic issues in interconnect layout stem from increased density of transistors that need to be connected, while economic issues pertain to reducing silicon (Si) real estate by stacking the metal interconnects.

This chapter examines the fabrication processes required for transferring product designs to Si. The main objective of this chapter is to present a simplified and logical flow for multilevel interconnect processing. Fabrication process technology must keep pace with constant advances in product designs for successful high-volume manufacture of new products. Such a developmental pace in process technology requires evaluation of existing process methodologies and development of new ones, often using a process technology driver such as a dynamic random-access memory (DRAM) device.

In the field of process technology development, new lithography processes are needed that are capable of extending current optical lithography limits and pushing them beyond the optical spectrum. Etching fine patterns in different substrates with severe restrictions on critical dimensions and on which bias are some of the requirements for plasma etching. On the dielectric deposition front, new types of insulators or combinations of insulators need to be further improved to provide conformal coverage of the device structures and fill ever-decreasing gaps between the metal interconnects. Dielectric materials are closely coupled with the planarization requirements and techniques. A complete global planarization is required due to lithography and metal step coverage constraints as the device geometries migrate to the levels of half-micron and below. Metallization requirements have become very significant as the interconnects occupy a significant portion of the Si real estate and have a strong influence on the overall reliability of the device.

12.2 Multilevel Interconnect System

A schematic diagram of a typical [Bartelink and Chiu 1989] three-level VLSI CMOS multilevel interconnect structure is shown in Figure 12.1, where several features of the multilevel interconnect system are shown.

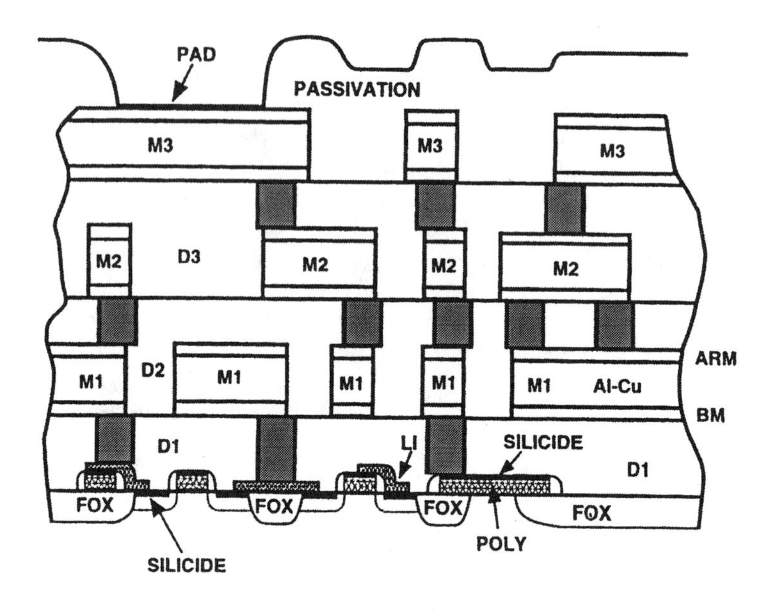

Figure 12-1 Schematic representation of a multilevel interconnect system [Bartelink and Chiu 1989].

In this diagram, the source/drain and polysilicon regions are silicided for reduction of contact resistance. Some processes use local interconnect (Li) or straps that may be formed as part of the silicide process to reduce device parasitics [Osinski et al. 1989]. Each dielectric layer is planarized to improve lithography and metal step coverage. Contacts and vias are opened in the dielectric films and filled with tungsten, W, as shown in Figure 12.1. The metal layers consist of a stack of films as each layer can effectively satisfy all the stringent requirements for interconnect metallization. A barrier metal layer (BM) is needed to prevent the aluminum (Al) from interacting with the adjacent layers. The main current-carrying conductor that is widely used is an Al alloy and several newer materials are in various stages of development. To reduce reflectivity of the metal layers, an antireflective material (ARM) is used. If such a layer is not used, the highly reflective metal surface will distort the ultraviolet (UV) light during resist exposure. This distortion leads to bridging resist and poor critical dimension control.

It is good practice to properly designate layers in a multilevel interconnect system. The whole purpose of the naming convention is to make the process flow convenient for manufacturing. Since there is no industry-wide standard nomenclature for a multilevel system designation, Table 12-1 shows how to designate various layers of a three-level metal

Table 12-1 Terminology of a Triple-Level Metal Interconnect

Type of Opening	Dielectric	Connecting Metallization	Line Metallization	Connecting layers
Contact	D 1/ILD 1	Plug 1	M1	Poly Si/Ml
Via 1	D 2/ILD 2	Plug 2	M2	M1/M2
Via 2	D 3/ILD 3	Plug 3	M3	M2/M3

system. This process can be extended to whatever depth and as many levels as required. Other examples of multilevel interconnect layer designations are available in the literature [Saxena and Pramanik 1984, Wolf 1990].

The first interlevel dielectric is between polysilicon and metal 1 and is usually a doped oxide. For dielectric layers, the designation could be D1, D2, etc., or ILD1, ILD2, etc., where D refers to the dielectric layer and ILD refers to the interlevel dielectric layer. In cases where more than one planarization is needed, planarization levels may be defines as Planar 1 or 2, etc. The openings in the dielectric layer from metal 1 to polysilicon and diffusion are referred to as contacts. Holes made through postmetal 1 dielectric layers are referred to as vias. In Table 12-1, W plugs are used to fill contact/via openings and are usually referred to as plug 1, 2, etc. an obvious designation. The metal layer consists of a stack of different metals, such as a barrier level, a shunt layer, a main conducting layer, and finally an antireflective layer. The metal layers that are used for interconnection as M1, M2, etc. are obvious.

12.3 Multilevel Interconnect Flow

The flow chart in Figure 12.2 shows the sequence of fabrication operations needed to manufacture an integrated circuit. This flow chart is the basis for discussion in this chapter and is used frequently. Specifically, the discussion will focus on the multilevel interconnect steps from dielectric deposition to metallization.

The first step in the fabrication is the deposition of a dielectric layer that isolates one level of metallization from the next. Since topography influences the step coverage of metallization layers, these dielectric layers are planarized. Once the necessary degree of planarization for a particular level of interconnection is obtained, the next step is to form contacts or vias by etching the dielectric to the underlying substrate. After this operation, a metallization step is used to fill the contacts/vias with appropriate metals for a proper electrical connection to the previous layer. Then metal interconnects are defined. This generic flow of events is repeated as many times as needed for fabricating devices with multiple levels of interconnects.

12.4 Dielectric Deposition

In the manufacture of integrated circuits, chemical vapor deposition (CVD) on dielectric is an extremely important function [Hey 1990, Gorowitz et al. 1987, Singer 1989, Burggraaf 1985, Blum 1987]. Dielectric isolation is essential to separate conduction layers from each other, and this isolation becomes more and more critical as the device geometries shrink and multiple layers of interconnects are used. In addition to isolation, dielectrics are used as a protective overcoat on the fully processed device. This protective overcoat protects the device from external contamination and scratches.

CVD dielectrics are used at different levels in integrated circuit (IC) fabrication processes. The deposition process varies depending on the thermal sensitivities of the underlying films and overall thermal budget for the device. Film properties largely depend on the CVD system employed and the chemistry used for deposition. There are three categories of commercial CVD systems commonly available [Singer 1989]. These are the low-pressure CVD (LPCVD), atmospheric-pressure CVD (APCVD) and plasma-enhanced CVD (PECVD).

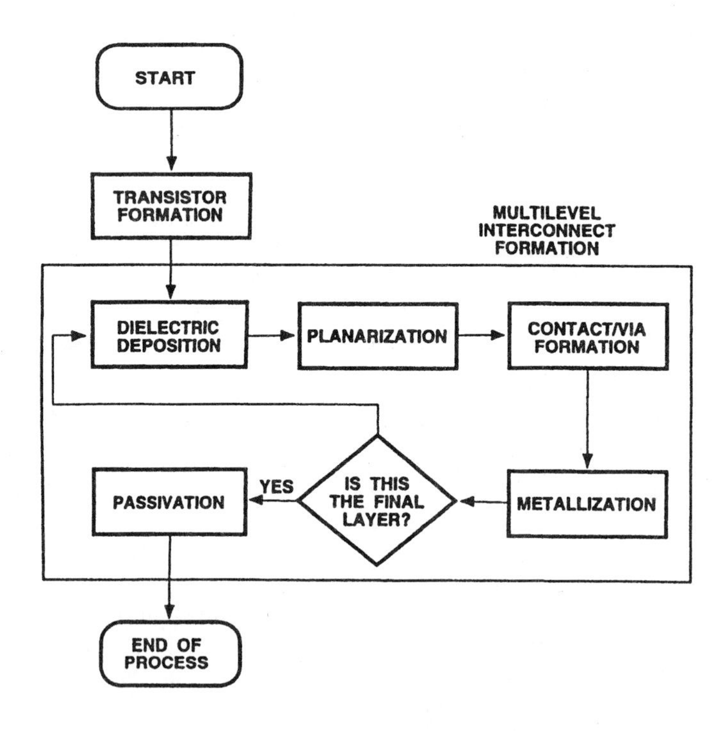

Figure 12-2 Processing sequence for multilevel interconnects

At the transistor level, CVD-deposited oxides and nitrides are used as sidewall spacers in lightly doped drain (LDD) structures. LDD structures are critical in metal oxide semiconductor (MOS) devices as they minimize the field strength close to the gate edge, thereby inhibiting injection of hot carriers. Once transistors are formed, an isolation layer is deposited that is usually an oxide doped with boron (B) and/or phosphorous (P). Contacts close to polysilicon and source/drain areas are formed in this film and filled with a conducting layer. Premetal dielectric deposition takes place at high temperatures, unlike the postmetal CVD oxides. Postmetallization CVD dielectric deposition offers a new set of challenges, as they have to be deposited at lower temperatures (< 400°C). These films are usually silane or tetrethylorthosilicate (TEOS)-based oxides and sometimes polymides are used as interlevel dielectrics (ILD). These ILD layers are subject to various planarization schemes in order to minimize the impact of topography on the next level of interconnects. The final protective overcoat or passivation layer is usually a combination of silicon nitride or oxynitride, oxide, and polymide films.

12.4.1 Applications

In very large-scale integration/ultra-large-scale integration (VLSI/ULSI) technologies dielectric materials are not used for various applications.

- For isolation of active areas in circuits
- Isolation of multiple interconnect levels
- Protective overcoat against moisture/impurities
- Sacrificial layer for planarization
- Alkali ion gathering
- Stress compensation
- Alpha particle shielding
- Transistor gates
- Etch mask (e.g., for trench etching)

12.4.2 Cause-and-effect analysis

The cause-and-effect diagram in Figure 12.3 illustrates the various influencing factors that affect the properties of the dielectric films. There are four basic factors that need to be considered when selecting and characterizing a dielectric film for use either as an interlevel dielectric layer or as a passivation layer. These are given below and each forms a separate branch in the cause and effect diagram.

Physical properties/defects. Physical properties of the dielectric layer are very important as they influence the films below and above it. Compared to thermal oxides, CVD oxides are inferior in quality and integrity. They exhibit higher levels of defects that may be particles, pinholes, and/or containment. The deposition parameters determine various physical and electrical properties of the film. The refractive index of a dielectric is a good monitor to evaluate the quality of the film. For example, the refractive index of silicon dioxide grown thermally is 1.46. A higher refractive index suggests a Si rich film and a lower value indicates a porous film. Porous films absorb moisture, which can cause damaging transistor threshold voltage shifts. A comparison of the refractive index of CVD oxide to thermal oxide gives an indication of relative quality. CVD deposition basically consists of introducing reactant gases, like silane (SiH_4) and oxygen (O_2), along with inert gases (for dilution) into the reaction chamber. In the reaction chamber wafers are arranged for uniform exposure to reactant gases. Once the energy (thermal or plasma) to initiate and sustain the reaction is supplied, reactant gases chemically react to form the dielectric film. Reaction can take place either on wafers or close to them. This results in uniform films with good quality.

In order for CVD film to isolate any two conduction layers, it must be able to cover the steps on which it is deposited. Problems in step coverage arise as metal spacing shrinks and metal thicknesses increase with each subsequent interconnect layer. The following analysis is valid for any film, dielectric or metal, being deposited over a step.

Figure 12.4 explains what step coverage is in relation to film thickness and Equations (12.1) through (12.3) show the relationship between step coverage and film thickness. Step coverage at a particular location is simply the ratio of the thickness at the location to the nominal film thickness [Wolf 1990, Ibbotson et al. 1988]. The aspect ratio is the ratio of the height to the width of the gap that must be filled. The higher the aspect ratio, the greater the difficulty in filling the gap.

$$SC_s = \frac{T_s}{T_t} \times 100 \tag{12.1}$$

$$SC_b = \frac{T_b}{T_t} \times 100 \tag{12.2}$$

$$AR = \frac{H}{W} \tag{12.3}$$

where SC_b is the bottom step coverage, SC_s is the side step coverage, T_s is the thickness at the thinnest point, T_b is the thickness at the bottom, T_c is the cusp thickness at the top, T_t is the thickness on flat surface, AR is the aspect ratio (H/W), H is the height, and W is the width.

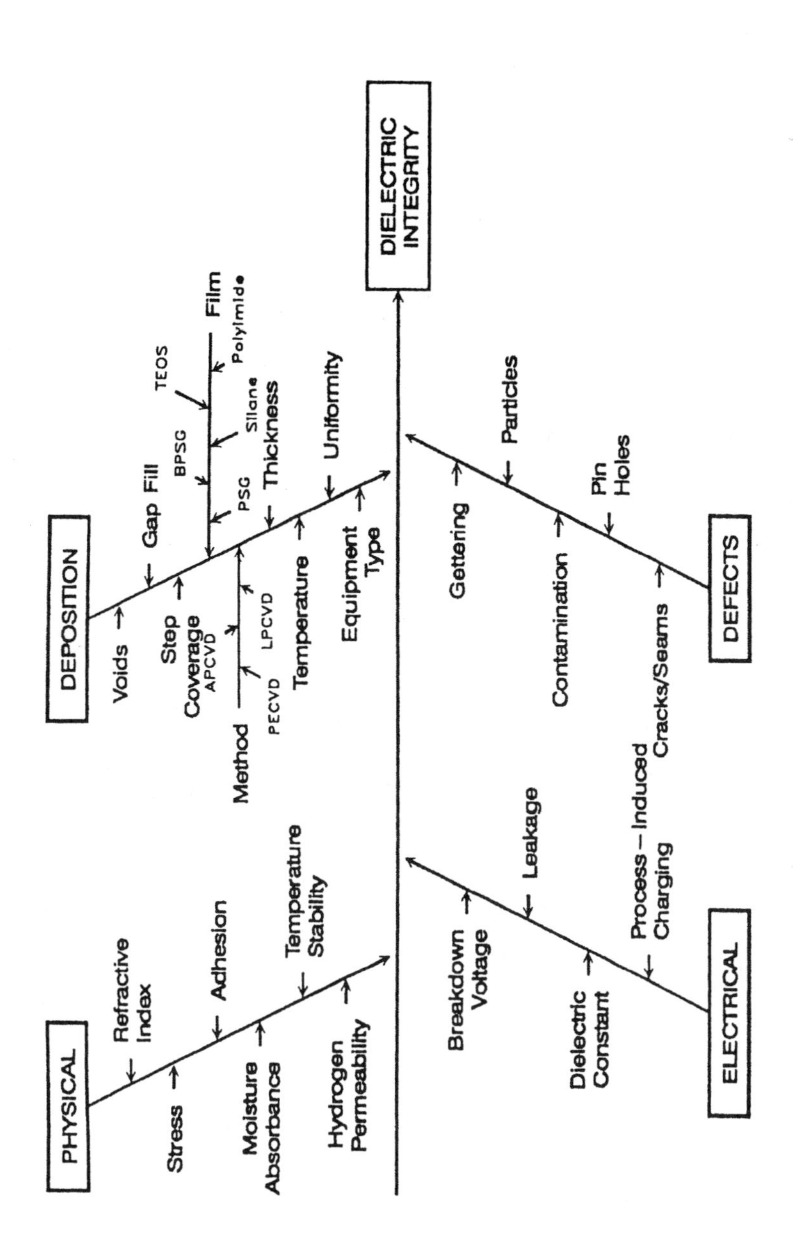

Figure 12-3 Cause-and-effect diagram for dielectric integrity

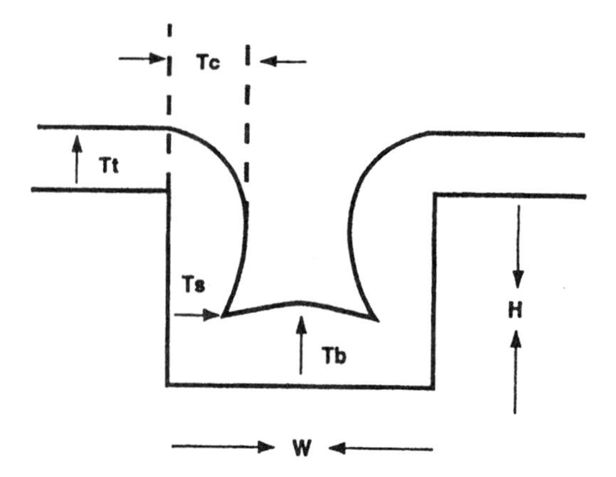

Figure 12-4 Step coverage analysis

A model [Raup and Cale 1989] has been proposed that explains the deposition and step coverage of plasma-enhanced CVD (PECVD) TEOS films.

This model assumes that activated atomic oxygen is responsible for silicon dioxide deposition. The explanation is that two reactions take place within the plasma that modulate the concentration of the atomic oxygen. Atomic oxygen is generated by ionization of oxygen molecules and depleted through volume recombination and chemical reaction on the surface. Table 12-2 summarizes the factors that influence step coverage and deposition rate. Optimum conditions should be determined by factorial experiments that offer a compromise solution between a high deposition rate and increased step coverage.

Another important CVD film deposition concern is the ability of the film to fill small page gaps without voids. Gap fill becomes an issue as the metal pitches become smaller (i.e., aspect ratios become larger). In

Table 12-2 Factors Influencing Step Coverage and Deposition Rate

Parameter	Step Coverage	Deposition Rate
F Power ↑	↓	↑
Pressure ↑	↓	↑
Temperature ↑	↓	↓

↑ is an increase and ↓ is a decrease.

conformal CVD deposition, cusping usually occurs as the film deposition takes place simultaneously from both sides of the trench, as shown in Figure 12.5. The cusping of the film over steps can be defined as follows:

$$C = \frac{T_C - T_S}{T_S} \times 100 \quad (12.4)$$

where C is the percentage of cusp, T_c is the thickness (cusp), and T_s is the thickness (side).

With continued deposition the leading edges of the cusp meet and form a void as further deposition takes place over the void. The voids may cause problems during planarization and etching as these voided regions are highly stressed and trap moisture/solvents that could out-gas under high vacuum and temperature. In order to eliminate the void formation, some commercially available CVD systems employ a deposition/etch process. After depositing a layer of oxide, the film is etched back and then the dielectric is redeposited. The etchback process causes a spacer formation between the steps in the gaps, and reduces the aspect ratio. This mechanism is further discussed in the TEOS oxides section. Further depositions and etches eliminate the voids. This is required for submicron gap filling as the aspect ratios increase with downward scaling of device geometries. A schematic representation of void formation in the dielectric due to high aspect ratios is shown in Figure 12.5.

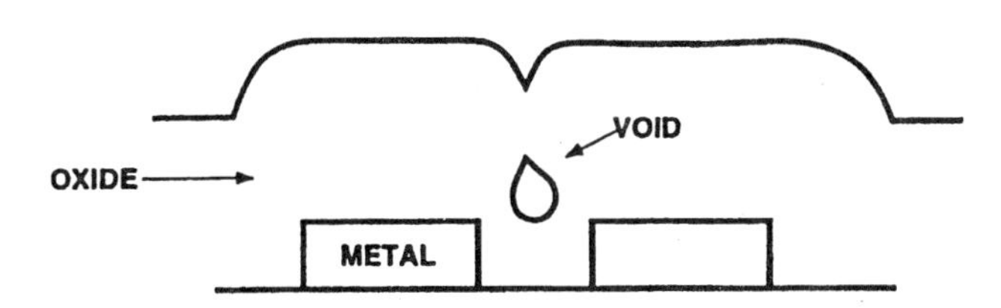

Figure 12-5 Schematic representation of void formation during dielectric deposition

Electrical properties. Understanding the electrical properties of the film is very important because they dictate the compatibility of the dielectric layer with the conducting films and also with other CVD films. The main purpose of the dielectric layer is to isolate active areas and conducting layers from one another. Although the dielectric layers do not conduct any current because of their insulating nature, they can modulate the electrical performance of the integrated circuit if their quality degrades. Dielectric films can induce spurious and unwanted charging on the transistor, which has a detrimental effect on the device reliability [Lefshitz and Smolinsky 1989, Hey et al. 1990, Kim and Puttlitz 1985]. The main electrical properties of interest for any dielectric layer are discussed below.

Capacitance. As the CVD dielectric film is sandwiched between two conducting layers, the capacitance created modulates the switching speeds of the devices. Voids that are formed as a result of the deposition process in narrow spaces could alter the dielectric constant of the insulator between the metal lines as they may trap moisture and other contaminants. The capacitance is a function of the thickness and dielectric constant of the film and both these parameters are considered when designing interconnect layouts. The following equation expresses the relationship between the capacitance of the insulator to its thickness and dielectric constant.

$$T_{ox} = \frac{\in A}{C} \qquad (12.5)$$

where T_{ox} is the thickness of the oxide, $\in$ is the dielectric constant of the oxide, A is the area of the capacitor, and C is the capacitance.

In order to ensure good dielectric integrity, films with higher dielectric constants are desirable.

Breakdown voltage/leakage. The breakdown voltage of an insulator is a good measure of its quality and its strength in withstanding an electric potential difference across it. Typical [Pramanik 1988] breakdown voltages for insulating films should be greater than 5 MV/cm. Oxide breakdown is an irreversible destruction of the dielectric. However, leakage or tunneling may not necessarily be destructive.

Process-induced charging. Process-induced charging is a problem that is not only affected by ion implantation and plasma etching/cleaning, but also by PECVD [Lefshitz and Smolinsky 1989, Hey et al. 1990, Kim and Puttlitz 1985]. In PECVD, energy to initiate the chemical reaction of reactant gases is primarily due to the ionization of the gases in a plasma. These ions can induce charging in the gate regions with high, direct current (DC) bias, if the gates are floating and do not have adequate charge dissipating structures/mechanisms in place. The charging is a function of the magnitude of the RF power (increased ion bombardment) and its ramp rate as the power modulates the plasma potential and the energy of the ions. If deposition is started with a high radio frequency (RF) power value, the process charging susceptibility of the device is higher than a slow, steady increase in RF power. In addition to this, some commercially available systems employ a sequence of deposition and etch to improve gap fill properties of the CVD film. The etch portion of this sequence may also contribute to the problem of charging. The correlation between power ramp and charging is still not well understood and needs to be examined more closely.

12.4.3 Pre-metal 1 dielectric BPSG

A dielectric layer is needed between polysilicon and the first metal for insulating polysilicon interconnects from metal interconnects. Current technologies use a metal silicide to lower contact resistance and these films are susceptible to degradation under high temperature conditions. Therefore, borophosphosilicate glass (BPSG) processing and its reflow must not degrade the silicide layers or alter the junction depths of the transistors. [Penning de Vries and Osinski 1989, Burmester et al. 1989]. In order to satisfy this requirement, BPSG has replaced phophosilicate glass (PSG) as an interlevel-dielectric material. PSG films require high temperatures of around 1000°C for reflow and also require a high concentration of phosphorus to improve the flow. The high phosphorous concentrations cause a whole new set of problems, for example, metal corrosion. To decrease the reflow temperature of the PSG films, the oxide is doped with a boron compound. Lowering of the temperature is achieved by addition of boron which makes BPSG quite compatible with the current submicron technologies. The primary function of the phosphorous in BPSG is to protect transistors from mobile ion contamination such as sodium. It has been shown [Kern and Smeltzer 1985] that BPG films getter (or trap) NA^+ and other mobile ion contaminants similar to PSG films.

Deposition. BPSG films can be deposited by CVD processes that are based on atmospheric pressure CVD (APCVD), low-pressure CVD film that is formed by reacting silane (SiH_4), phosphine (PH_3), diborane (B_2H_6), or boron trichloride (BCL_3) and O_2 diluted with an inert gas, usually nitrogen (N_2), at the surface of the wafer. Control and management of gas flow ratios and deposition conditions are important for obtaining the desired film thickness and dopant concentrations. BPSG is formed by the combined oxidation of the various hydrides. The oxidation kinetics are different for BPSG as compared to the oxidation kinetics of any single hydride. Table 12-3 summarizes the impact of varying several process parameters such as flow rates and dopant concentrations, etc. on the weight percent of boron and phosphorous and the film deposition rate. Nitrogen is used as a diluant for the hydrides. In the case of APCVD, BPSG deposition exhaust pressure is one of the key parameters. The greatest impact on B and P concentrations is obtained from modulation of concentration and flow rate of their respective

Table 12-3 Interrelationship between BPSG Deposition Parameters

Parameters	Wt. % B	Wt. % P	Deposition rate
$[B_2H_6]$ ↑	↑↑	↓↓	—
$[PH_3]$ ↑	—	↑↑	—
O_2:hydride ↑	—	—	↑
B_2H_6 flow ↑	↑↑	↓	↑↑
PH_3 flow ↑	↓	↑↑	↑
N_2 flow ↑	↓	↓	↓
Deposition temperature↑	—	—	↓
Pressure ↑	↓	↓	↓

↑ is an increase; ↑↑ is a large increase; ↓ is a decrease; ↓↓ is a large decrease; and – has no significant impact.

hydrides. Selection of the boron and phosphorous weight percent for a particular device depends on its thermal budget and planarization requirements.

The relationships in Table 12-4 are observed and monitored in a typical BPSG process that is used for device fabrication [Kern and Schnable 1982]. When the PH_3/SiH_4 ratio increases, the phosphorous concentration in the film also increases, but has little effect on the boron concentration. However, the interaction is different when the B_2H_6/SiH_4 ratio is increased. This causes a significant increase in the B concentration, but also causes the P concentration to decrease. The reason for this may be explained by the reaction kinetics of the oxidation of the hydrides. The oxidation of the boron hydride proceeds much faster than oxidation reactions of both silane and phosphine as diborane reacts more readily with oxygen. The oxidation reactions [Tong et al. 1984] of the three hydrides are as follows:

$$B_2H6 + 6N_2O \rightarrow B_2O_3 + 6N_2 \quad (12.6)$$

$$SiH_4 + 4N_2O \rightarrow SiO_2 + 4N_2 \quad (12.7)$$

$$aPH_3 + bN_2O \rightarrow cP2O_3 + dP_2O_5 + eH_2O + fN_2 \quad (12.8)$$

where coefficients *a* through *f* are unknown.

Densification. As the contacts are defined in the BPSG film, the film integrity and flow characteristics are critical for contact lithography and etching. In 1 μm and larger technologies, the densification is done following the deposition of film and then the contacts are defined. Reflow of contacts after contact formation changes the profiles of the contacts and helps improve the metal step coverage. In submicron technologies control of contact dimensions becomes difficult if reflow is done after the contacts are formed. These technologies rely on other contact planarization techniques and not on the BPSG reflow process. Therefore, the densification and reflow may be combined and performed in one single step.

During oxidation of the hybrids of B and P their respective oxides are formed. The as-deposited film contains these oxides and is amorphous in nature. The densification process must take place soon after deposition, otherwise these oxides would react with the moisture in the ambient and form crystals. The densification of the BPSG film in steam at around 800°C causes hydrolysis of the oxides and also decomposition and vaporization of some of them.

In the BPSG film, P is present as P_2O_5 and P_2O_3 oxides [Tong et al., 1984]. Hydrolysis of these oxides results in formation of their respective acids. Boron oxide also reacts with water to form boric acid. This reaction can take place at room temperature and may result in boric acid crystal formation. This reaction is given below.

$$B_2O_3 + 3H_2O \rightarrow 2H_3BO_3 \tag{12.9}$$

Acids formed by hydrolysis of both boron and phosphorous oxides react further and may cause formation of BPO_4. The impact of these byproducts on the device defect density is discussed below.

Reflow. The purpose of reflow is to planarize the BPSG film so that the underlying topography may be transformed to a gentle and smooth surface. The BPSG flow angle decreases as the reflow temperature increases. However, the reflow temperature is constrained by the device channel length specifications as an increase in temperature may alter the source/drain profiles. In addition, silicided polysilicon and source/drain areas are sensitive to increases in temperature as the silicide will become unstable.

It has been reported [Tong et al., 1984.] that for a given BPSG composition, achieving equivalent flow angles in both nitrogen and steam ambience requires a 70°C or higher temperature in the dry nitrogen ambient. The steam ambient leaches the dopants from the top surface of the BPSG film. This has beneficial effects, as the subsequent metal films would contact a lower doped oxide. Steam also oxidizes the interface of BPSG and silicon substrate to form a thin layer of oxide that prevents autodoping of the wells.

Flow and defect characteristics. Success of the reflow of the BPSG and its impact on planarization is measured by the flow angle. Flow angles are not only dependant on film deposition conditions and reflow parameters, but also on the underlying topography. BPSG film thickness tends to be larger over dense topography as compared with isolated structures. The reflow process contributes to thickness variation by making the BPSG flow which smoothes the topography by depletion of the film over steep structures and accumulation in the gaps between polysilicon interconnects. This presents a situation of different contact depths during the contact etch process because the thickness of BPSG varies over dense and isolated structures. The scanning electron microscope (SEM) micrograph in Figure 12.6 illustrates the flow characteristics of BPSG over steps with varying spacing. To improve the planarization and obtain lower flow angles, boron and phosphorous concentrations need to be increased.

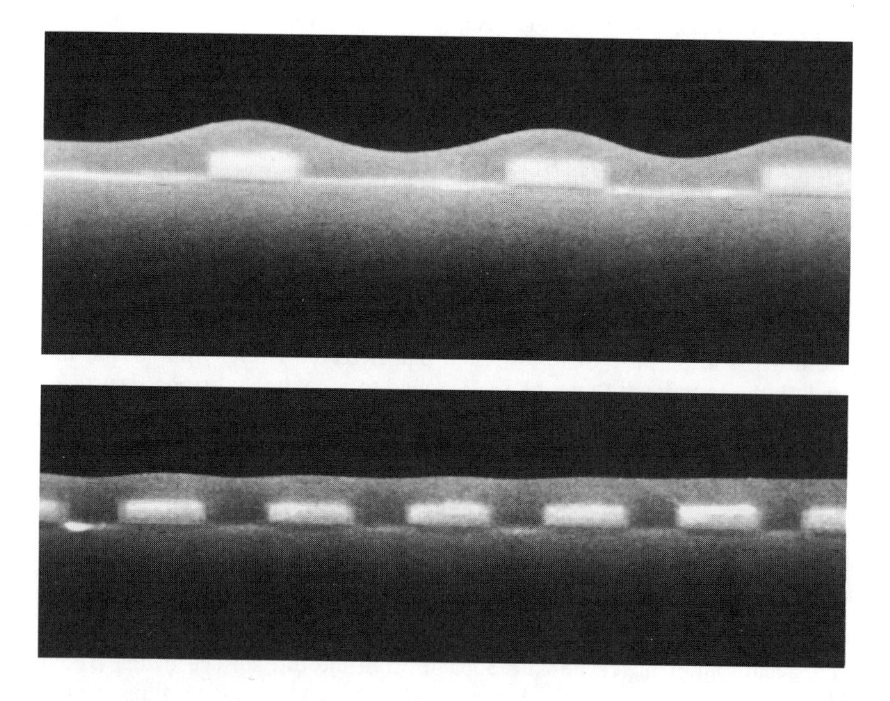

Figure 12-6 Pre-metal 1 planarization-BPSG over topography

However, this increase may have a detrimental effect on yield and the reliability of the device due to the formation of BPSG crystals. There are two types of crystals that are specific to BPSG processing. These are boric acid and borophosphate crystals. Boric acid crystals form at room temperature and are quite easily soluble in water and are usually removed in the stream densification/reflow cycle. In order to control the formation of crystals, the time delay between the BPSG deposition and densification/reflow must be minimized. The boric acid crystals are quite large, between 5 and 10 μm in size. The second type of defect is the formation of the borophosphate (BPO_4) crystals which are not water soluble. These crystals are formed during steam reflow when the reflow parameters are less than optimal. BPO_4 crystals are small in size, usually less than 1 μm.

In order to minimize crystal formation, a lower concentration of dopants is preferred, but such action may lead to higher flow angles. The relationship between the dopant 0 concentration, flow angles, and crystal formation is illustrated in Figure 12.7. This presents an interesting problem – lower flow angles are needed to enhance the planarization but at the same time lower defect densities (due to crystal formation) are needed to prevent yield degradation. Such constraints force this process to operate within a very narrow process window.

(Some materials presented in this section are from the following references: Hurley et al. 1987, Penning de Vries and Osinski 1989, Becker et al. 1986, Hurley 1987, Kern and Smeltzer 1985, Kern and Kurylo 1985, Burmester et al. 1989, Johnson et al. 1987, Mercier 1987, Kern and Schnable 1982, Carpio 1989, Tong et al. 1984, Kern and Smeltzer 1985.)

12.4.4 Post-metal dielectric CVD oxides

The post-metal 1 dielectric deposition process comes with its own stringent requirements for submicron devices. Note that for post-metal 2 or 3, etc. similar process methodology is used; the dielectric thickness may vary from layer to layer due to metal thickness differences between one level and the next.

CVD technology that has been used up to about 1.0 μm design rules has not been based on the oxidation of silane [Riley et al. 1989]. Silane oxide does not deposit conformally over underlying topography and hence is not adequate in filling gaps, the aspect ratios of 0.5 or larger, for submicron technologies [Ting 1991]. TEOS-based oxides have improved the conformality of the deposited oxides. Improved conformality helps in preventing voids in thick films and also presents a quasiplanarized surface. But is TEOS deposition process adequate for sub-half-micron technologies? This section examines the relative advantages of TEOS-based oxides over silane oxides and the process integration issues involved in incorporating a dielectric film that satisfies both the physical and electrical requirements for submicron devices.

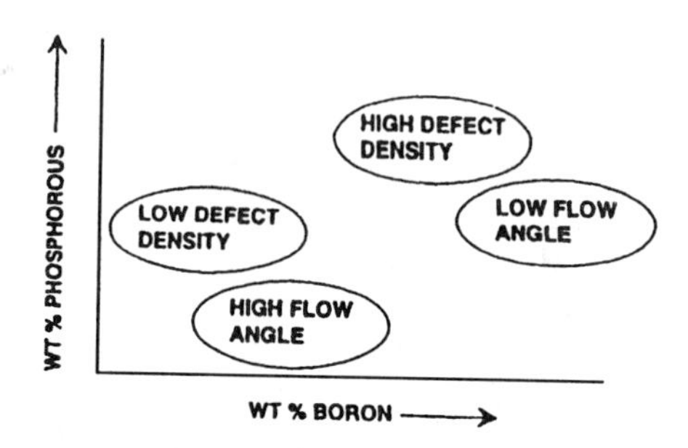

Figure 12-7 Impact of B and P on defect density and flow angle of BPSG films

Silane oxides. Oxides formed by the oxidation of silane have been limited in their use because of their deposition characteristics, especially for submicron process technologies with high aspect ratios [Riley et al. 1989, Magnella and Ingwersen 1988, Pliskin 1977, Adams et al. 1981]. Silicon dioxide is formed by the reaction of SiH_4 and O_2 in a low pressure regime of less than three torr and around 400°C. This reaction takes place at the wafer surface and is primarily governed by the rate of arrival at the surface of the reactant species from the gas phase and, of course, the temperature of the surface [Riley et al. 1989]. To achieve a conformal coverage of steps or fill gaps, the surface mobility of the active species must be high and the reaction must be uniform at the surface. It has been observed [Riley et al. 1989] that during the deposition of silane-based oxides, the active species have very little surface mobility, possibly less than 1 μm. The surface mobility could be enhanced by increasing the wafer temperature, but this poses a major problem with the metal layers underneath the interlevel dielectric. In order to prevent the aluminum interconnects from voiding and to suppress hillock formation, the temperature must be low.

Since silane-based oxides suffer from poor surface mobility, the deposition of the oxide in narrow gaps and over wide structures varies and is a function of the arrival angle of the reactive species from the gas phase. In wide-open areas, the arrival angle is high and therefore the deposition rate is also high. On flat wafers the deposited oxide is quite uniform. However, in narrow spaces the arrival angle is a function of the aspect ratio. The higher the aspect ratio, the smaller is the arrival angle. The amount of reactant species arriving and reacting to form SiO_2 in high-aspect ratio spaces is less over open and wide structures. This causes localized differential deposition rates, probably due to gas depletion. Therefore, for a given film thickness, the open areas will reach the required thickness before the narrow gaps. As the deposition continues over time, the film gets thicker and voids begin to form in narrow spaces. The void formation is due to poor surface mobility and deposition rate variations.

Silane chemistry is adequate for technologies above 1 μm. For submicron technologies this technique of ILD deposition may not be adequate if used by itself. Silane chemistry adapted for submicron technologies by using a multilayered dielectric layer in combination with spin-on-glass (SOG) has been used quite successfully [Yen and Rao 1988]. A thin layer of oxide deposition followed by SOG can be used either in an etchback or a nonetchback mode. Details on SOG planarization are given in the appropriate section. Such a dielectric stack allows sufficiently thin oxide film to be deposited so the problem of void formation is eliminated. Gap fill is achieved by the SOG film. From a safety point of view, silane gas delivery and exhaust management is very important as silane is very flammable.

TEOS oxides. TEOS-based oxides [Hills et al. 1989, Kotani et al. 1989, Emesh 1989, Normandin 1989, Kamins 1978, Becker and Rohl 1987, Hey and Machado 1989, Thomas et al. 1987, Wu et al. 1988, Nguyen et al. 1990, Ahlburn et al. 1991] deposited at low temperatures (<400°C) have been developed as an alternative dielectric to silane-based oxides because they tend to have better gap fill properties [Magnell and Ingwersen 1988]. Here again, the surface mobility is believed to play a pivotal role in enhancing the step coverage and also the gap fill. The reactant species in TEOS-based oxides tend to have a higher surface mobility than the silane-based oxides. In addition to a higher mobility, the TEOS reactant species exhibit a decrease in sticking coefficients.

TEOS silicon dioxides may be deposited in a plasma-enhanced mode where the TEOS reacts with oxygen in a plasma. This type of oxide may be designated as plasma-enhanced TEOS (PETEOS). PETEOS silicon dioxide has better conformality than silane based oxide, but it is not perfect. Typical conformality values are less than 75 percent in narrow spaces [Ting 1991]. Silicon dioxide may also be formed by LPCVD reaction of TEOS with

$$Si-(O-C_2H_5)_4+O_3 \rightarrow SiO_2+by-products \qquad (12.10)$$

ozone at temperatures of less than 400°C. Oxide formed by this technique is termed thermal TEOS (TTEOS). The TTEOS reaction proceeds as follows, even though this reaction is a result of several intermediate reactions.

TTEOS silicon dioxide films are nearly 100 percent conformal but suffer from being very porous and absorbing a significant amount of moisture [Nguyen et al. 1990]. To further enhance the step coverage and film integrity of TTEOS oxides, the deposition pressure may be varied. TTEOS/O_3 may react at atmospheric or subatmospheric pressures. The change in pressure causes the films to be denser and absorb less moisture. For the 16-megabyte (MB) DRAM process a subatmospheric (600 torr) TTEOS deposition has been used to fill trenches 0.5 μm wide by 1.0 μm deep [Ahlburn et al. 1991].

The current strategy to fill narrow spaces without voids is to use TEOS-based oxides and simultaneously decrease the aspect ratio of the gaps so that the dielectric material can fill the spaces easily. But how can we decrease the aspect ratio? A deposition/etch method [Pennington 1989] may be used to improve the deposition characteristics using TEOS-based oxides. Depositing a specified thickness of oxide and then etching it back partially creates a spacer in the gaps. The spacer changes the aspect ratio and further deposition of oxide can be done without void formation. The spacer formation may be achieved by either argon (Ar) sputter etching of the oxide and/or reactive ion etch (RIE) etchback of the oxide. Deposition, etchback, and redeposition are all done in one machine, but in different chambers. Generally, the PETEOS and TTEOS are used in a deposition/etchback dielectric deposition scheme where the TTEOS is used only for gap filling.

After the TTEOS is used to fill the gaps, through single or multiple depositions and etchbacks (AR sputter etching or RIE), PETEOS oxide or other oxides are redeposited for planarization. Figure 12.8 shows a schematic illustration of the TTEOS spacer sandwiched between two PETEOS layers.

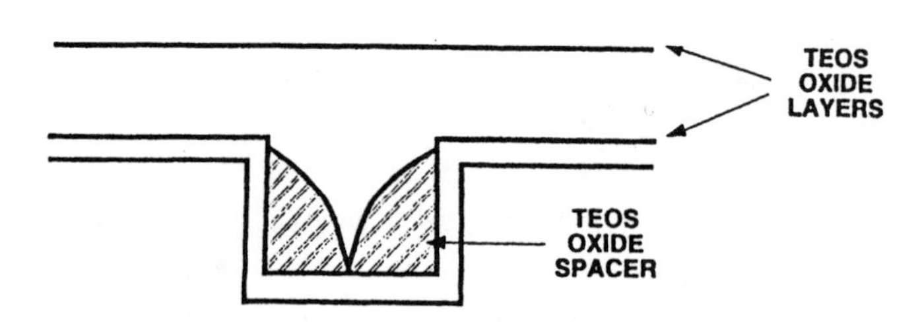

Figure 12-8 TEOS-based oxide deposition in narrow spaces

PETEOS is a denser and drier film than TTEOS because the reaction of TEOS and oxygen in a plasma is more efficient in dissociation and enhancement of the reactant species in this gas phase oxidation. Furthermore, the plasma assists the surface reactions through ion bombardment. This results in PETEOS films having higher deposition rates, lower hydrogen concentration (therefore a drier film), improved bonding structure (denser film), and overall better film properties than TTEOS. One of the drawbacks of using TEOS-based oxides is their effect on the transistor threshold voltage stability. Threshold voltage shifts can occur when moisture diffuses from the dielectric layer to the gate region where it creates electron traps.

12.5 Dielectric Planarization

This section deals with planarization of dielectric layers and follows the dielectric deposition module as shown in Figure 12.2. Planarization has become an integral part of the multilevel interconnect scheme where topographical interferences with fine geometry patterning and etching pose a problem [Saxena and Pramanik 1986, Nagy and Helbert 1991, Kup 1987, Ting 1991, Moghadam 1992]. The source of this problem lies in the fact that several levels of metal interconnects are required to fabricate integrated circuits with millions of transistors per chip. The current distribution requirements for the different metal interconnects demand appropriate film thicknesses for both the metal and dielectric films between them. A stack of several layers of metal separated by insulating films make the topography very uneven with extremely large steps. Planarization is essential for the following reasons.

- *Step coverage.* The step coverage of metal impacts the current-carrying integrity of the interconnects. Poor metal step coverage will cause current crowding and will lead to electromigration failures. Planarization is required for multi-level metal technologies and beyond.
- *Differential bias/critical dimensions.* Planarization, along with the antireflective layers, prevents spurious UV light reflections from the sides of metal as it goes over steps. Such reflections can cause loss in critical dimensions and also cause defects in the metal line. Planarization is required to overcome differential bias due to variations from the sides of metal as it goes over steps. Such reflections can cause loss in critical dimensions and also cause defects in the metal line. Planarization is required to overcome differential bias due to variations in resist thickness on unplanarized surfaces.

12.5.1 Requirements

There are basically two requirements for any planarization process.

- *Uniformity.* Uniformity is an important output parameter that needs to be characterized and monitored to ensure that all the dielectric is planarized irrespective of the circuit density. One of the key uniformity parameters is the within-a-die (WID) uniformity. This determines if the planarization is *local* or global in nature, which will be described in this section. Within-a-wafer (WIW) uniformity is also an important parameter as the via etch uniformity is influenced by it.

- *Manufacturability.* The planarization process has to be manufacturable. This means the machine throughput must be reasonably high to support high-volume production. Under the conditions of high output the defect density must be very low as yield and reliability must not degrade.

12.5.2 Local and global planarization

Planarization can be local or global in nature [Sivaram et al. 1991, Technical note Westech Systems Inc., Daubenspeck et al. 1991]. *Local planarization* refers to planarization in dense regions of the device where planarization [Technical note Westech Systems Inc.] occurs over submicron geometries with pitches varying from less than 2 to about 4 μm. *Global planarization* refers to planarization over a large distance of about 150 μm or greater [Technical note Westech Systems Inc.]. In some planarization procedures, it is essential to optimize for local planarization and sacrifice global planarization. This is a problem that is usually faced in the etchback type of planarization processes. For example, memory circuits, such as DRAMs, have a very dense concentration of structures in the array as compared to the periphery of the die. Therefore, they need an array-optimized planarization process. However, for microprocessors and other logic devices, planarization should be optimized evenly across the die as the structure layout is not as systematic and symmetrical as in the memory circuits.

12.5.3 Summary of planarization technologies

There are several dielectric planarization processes currently being used for 1 μm and submicron process technologies.

Resist etchback. [Daubenspeck et al. 1991, Wison 1986, Fritzshe et al. 1986, Riley and Castel 1987]. This is a very common and convenient method for dielectric planarization. Planarization of the oxide surface is achieved by coating the wafer with a sacrificial resist layer and then etching it back with approximately equal oxide and resist etch rates. After resist is spun on the wafer, the surface is planarized because the resist fills the depressions and is thin over steps. During plasma etchback, the oxide over the high points gets etched before the oxide in the spaces. The etching is continued until the desired oxide planarization is obtained.

SOG (with and without etchback). [Gupta 1988, Tompkins and Tracy 1990, Molnar 1989, Pai et al. 1987, Watanabe and Todokoro 1988, Ting et al. 1987, Whitwer et al. 1989, Yen and Rao 1988, Shacham-Diamand and Nachumovsky 1990, Vollmer et al. 1988, Schiltz 1986, Kojima et al. 1988, Gupta 1989, Kawai et al. 1988, Woo et al. 1990, Ito et al. 1990,

Lifshitz et al. 1989, Parekh et al. 1987]. There are two methods of using SOG for dielectric planarization. The first is a nonetchback process. This involves spinning, baking, and curing the SOG until the desired planarity is obtained. Additional oxide is deposited over the SOG to meet the dielectric thickness specifications. The second method is an etchback process where SOG is used as a sacrificial layer, similar to the resist etchback process.

Chemical mechanical polish. [Sivaram et al. 1991; Rentelin et al. 1990 Technical note Westech Systems Inc.; Davari et al. 1989; Thomas et al. 1990; Patrick et al. 1991; Technical note Westech Systems Inc; Comello 1990; Warnock 1991]. Planarization is achieved by polishing the dielectric surface on a rotating abrasive pad. This causes the high spots on the dielectric surface to be polished before the low spots. The chemical component is provided by the slurry which has two primary functions. First, it serves as a lubricant, and second, it helps break the SiO_2 bonds at the surface.

Bias sputtered quartz. [Singh et al. 1987]. The wafers are planarized by sputter deposition and etch of dielectric material from a target onto the wafer. During the sputtering process, the wafers are held at a negative potential relative to the plasma and the bias voltage may be varied to obtain the desired planarization. In this process, sputter etching and deposition are simultaneously taking place. As the material is being deposited, the high points on the wafer surface are sputter etched. This dual action planarizes the surface.

Thermal flow. Planarization using thermal flow is restricted to the dielectric layers over polysilicon and before metal layers are deposited. The oxide is doped with B and P to improve the flow characteristics and to lower the flow temperature. This process was discussed in detail above in Section 12.4.

Deposition/etch. Good conformal coverage may be obtained by using a LPCVD/PECVD deposition/etch process. The material is deposited and etched in sequence a few times until the gaps between metal lines are completely filled. In addition to the deposition/etch sequence, the use of TEOS-based oxides has also helped in achieving good gap fill and planarization.

Table 12-4 is a comparison of some planarization processes. The properties of the planarization processes that have been compared are the following.

Table 12-4 Comparison of Planarization Schemes

Planarization Process	Local	Global	Feature size (μm)	Mfg
SOG	Yes	Some	1	Good
Etchback	Yes	Some	1	Good
CMP	Yes	Yes	<1	Fair
Thermal Flow	Yes	No	<1	Good
Deposition/etch	Yes	No	<1	Good

- *Local/global planarization.* It is important to recognize the primary benefit of various processes. Some have superior local planarization properties and others are suited for a global planarization.
- *Feature size.* The feature size dictates the process technology for which the planarization step is best suited.
- *Manufacturability (Mfg).* This aspect indicates the adaptability of the planarization process in a high-volume manufacturing environment.

12.5.4 Degree of planarization

We can evaluate the success of any planarization technique by measuring its degree of planarization and the planarization distance. The degree of planarization, DP, is a measure of the dielectric step coverage after planarization and is defined by Equation (12.11):

$$DP = 1 - \frac{(SH)_{post}}{(SH)_{pre}} \tag{12.11}$$

where DP is the degree of planarization and SH is the step height.

If DP approaches 1, that means the pre-step height is very large as compared to the post-step height. In such a situation the planarization is perfect over long ranges. If the pre- and post-step heights were to be equal, then the DP would be equal to zero. This means there is no long-range planarization.

Another figure of merit that is indicative of the planarization effectiveness is the planarization distance (PD). Planarization distance is a measure of the wafer smoothness over long distances and is the distance of the step in topography after planarization. A long PD implies that the variations in topography smaller than the PD will be completely planarized. It is not surprising to find PDs up to a few millimeters for a chemical mechanical polish planarization process. As compared to CMP, SOG planarization may yield a PD of about 2 μm [Sivaram et al. 1991]. Higher PD values are indicative of global planarization.

Instead of describing each of the different planarization processes, the focus of this section is to examine the issues pertaining to local and global planarization. To illustrate each of these, a planarization method has been selected that best represents either local or global planarization. For local planarization, SOG process with and without etchback is discussed. For global planarization, chemical mechanical polishing is examined.

12.5.5 Global planarization (CMP)

Chemical mechanical planarization (CMP) is gaining wide acceptance as it is the only process to date that can offer better global planarization for VLSI/ULSI devices than the other planarization processes [Sivaram et al. 1991, Rentelin et al. 1990, Technical note Westech Systems Inc., Davari et al. 1989, Thomas et al. 1990, Patrick et al. 1991, Technical note Westech Systems Inc., Comello 1990, Warnock 1991]. The use of CMP for planarization was first used [Davari et al. 1989] in combination with RIE to planarize CVD oxide for shallow trench isolation as part of the 16-MB DRAM process. This combination helped in reducing process nonuniformities associated with each process when used independently. There has been a lot of activity in the development of CMP since then. Using chemical mechanical polishing by itself to planarize interlevel dielectric oxides has been reported with very encouraging results [Sivaram et al. 1991]. Therefore, the discussion below focuses on planarization with CMP alone and not in combination with other techniques.

The CMP cause and effect diagram shown in Figure 12.9 illustrates the various parameters that affect oxide polishing. The CMP method may be briefly described as a process where the wafers are held face down against a rotating polishing pad. This pad is constantly lubricated by a slurry that consists of colloidal silica at a high pH. It is this mechanical action that enables the oxide to be removed from the surface. The polish process flattens out height differences, since the high areas of topography are removed faster than low areas. When this is accomplished, all traces of the original topography are removed and an acceptable post-polish surface is obtained. The CMP process can planarize both locally and globally over topography with varying pitches. Local planarization is shown in Figure 12.10; global planarization is demonstrated in Figure 12.11.

The two main components of the CMP process that remove the surface oxide are the mechanical and chemical actions. Both these are necessary for the polishing of the oxide surface and are discussed below.

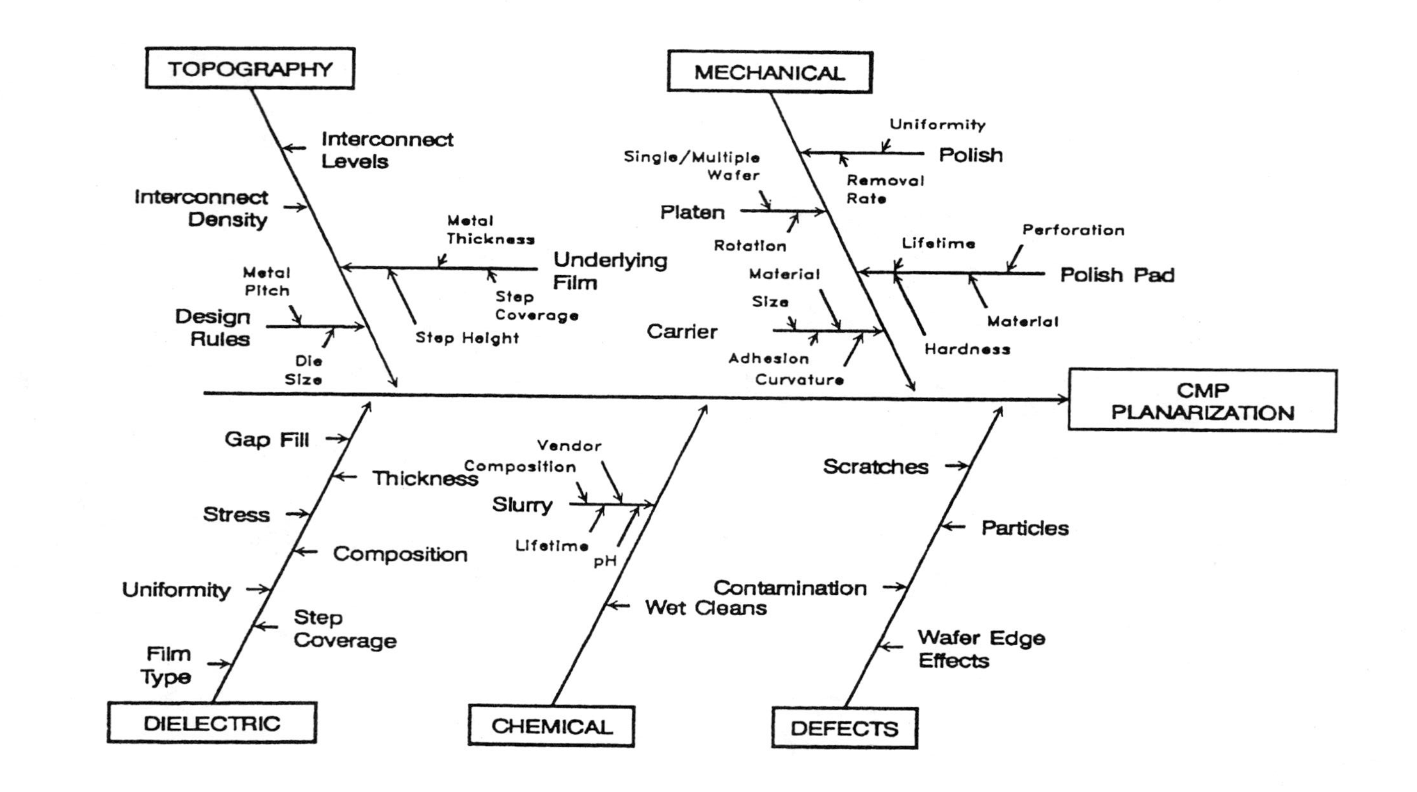

Figure 12-9 Cause-and-effect diagram for chemical mechanical process

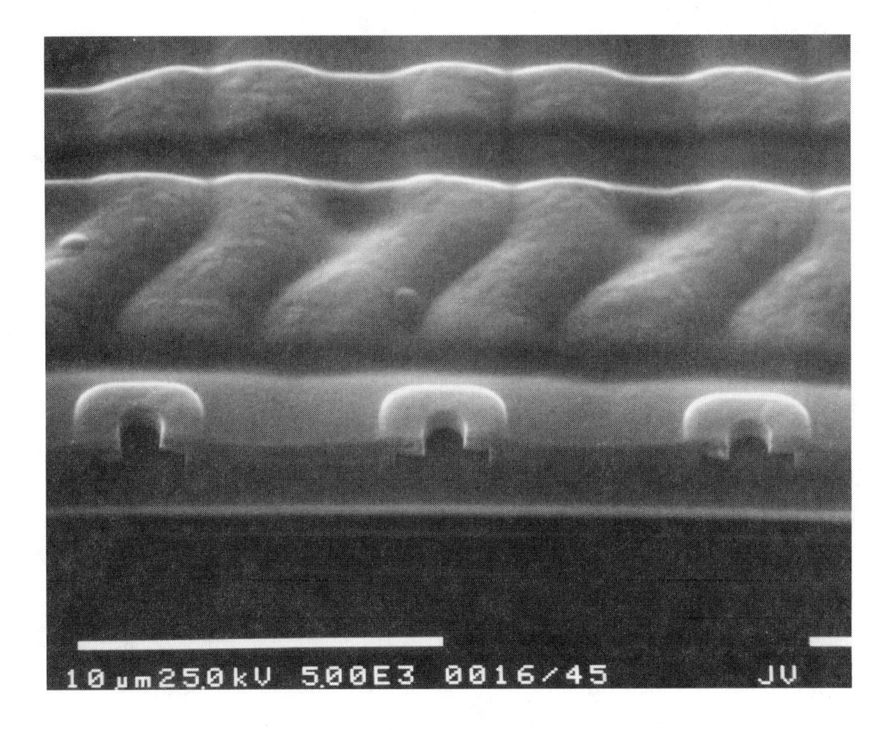

Figure 12-10 Example of local planarization

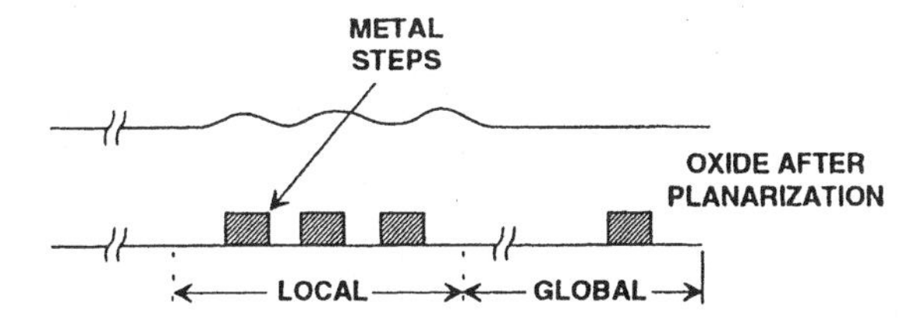

Figure 12-11 Schematic illustration of local and global planarization

Mechanical component. The mechanical action consists of a rotating table with a platen on which the polishing polyurethane pad is mounted. The wafer is pressed down against this polishing pad by a wafer carrier, whose shape may be adjusted to improve the polish uniformity. Downward pressure is then applied to the wafer. The hardness and aging of the polish

Table 12-5 Process Interactions in CMP

Parameter	PD	Removal rate	WIW
Platen Speed ↑	↓	↑	—
Down Pressure↑	↑	↑	—
Pad age ↑	↓	↓	↓
Pad hardness ↑	↓	↓	↓

pad affect the oxide planarization. The removal rate of the oxide continuously drops off as the pad ages and the perforations in the pad are blocked off and therefore prevent slurry distribution. For effective process control, the pad needs to be conditioned by constant and regular cleaning, usually through an abrasive process that opens the channels for the slurry to flow. This action restores the oxide removal rate. The removal rate of the oxide, from a mechanical viewpoint, is a function of the downward pressure, the relative velocity between the oxide surface, and the polishing pad, age/conditioning of the pad, and temperature.

Table 12-5 summarizes the interactions of the parameters that are part of the mechanical component of the polish process. This table shows the effect on the degree of planarization, the removal rate and the WIW uniformity. The up arrow indicates an increase and a down arrow a decrease in value of the parameter. When the effect on the output parameter is negligible, it is denoted by a dash (–).

Chemical component. The mechanical polishing action is complemented by the chemical component in the form of slurry introduction. The slurry, consisting of suspended colloidal silica in potassium hydroxide, reacts with the dielectric film by forming a hydroxylated surface layer. It is thought that several surface interactions take place between the slurry particle and the oxide surface. These reactions result in the disassociation of the Si-O-Si molecule through the diffusion of water through the oxide. The bond strength between the slurry particle and the wafer surface determines the kinetic coefficient of friction between the two surfaces during polishing. A high pH is needed for the suspension of the silica and for the reaction to take place. The by-products of this reaction are removed by the mechanical action pad and the colloidal silica.

12.5.6 Local planarization (SOG)

SOG is being widely used and remains one of the most promising materials for interlevel planarization of VLSI multilevel interconnects [Gupta 1988, Tompkins and Tracy 1990, Molnar 1989, Pai et al. 1987, Watanabe and Todokoro 1988, Ting et al. 1987, Whitwer et al. 1989, Yen and Rao 1988, Shacham-Diamand and Nachumovsky 1990, Vollmer 1988, Schiltz 1986, Kojima et al. 1988, Gupta 1989, Kawai et al. 1988, Woo et al. 1990, Ito et al. 1990, Lifshitz et al. 1989, Parekh et al. 1987]. By comparison with other planarization techniques, the SOG process is a relatively simple technique for planarization surfaces with severe topography both below [Whitwer et al. 1989] and above the first metal layer. As SOG is similar to SiO_2 after its final cure, it can be made a part of the dielectric stack and can be easily integrated with CVD oxides to obtain stable insulating properties. This can be achieved by proper selection of SOG material and optimizing its coating, baking, and curing steps. Unlike CMP, COG does not offer complete planarization, instead it smoothes out the underlying topography. This results in a reduction in via depth differentials. Even though SOG is a simple planarization technique and has several advantages over other processes, [Schiltz 1986] the reliability implications of its use must be carefully examined. Moisture-induced transistor threshold voltage instabilities and other reliability issues are some of the major concerns with SOG planarization [Shacham-Diamand and Nachumovsky 1990, Vollmer 1988].

SOG is a liquid and thus can be easily spun on wafers. This action causes the SOG to fill gaps or depressions in the topography and provide a planar surface. Several thermal treatments are necessary to ensure that the desired thicknesses obtained and the solvents are removed from the SOG. To illustrate the planarization properties of SOG, Figure 12.12 presents a generic relationship between the degree of planarization and the metal line spacing for two different metal thicknesses, H and 2H. In narrow spaces, SOG fills the gaps and is thick in those areas. As the metal line spacing is increased and the gaps widen, SOG thickness in these areas decreases and consequently, the degree of planarization decreases. The SOG cause and effect diagram given in Figure 12.13 provides an overview of the SOG planarization process. This diagram highlights some of the salient aspects of SOG planarization, which are discussed below. Key to the successful implementation of a SOG planarization process lies not only on the technique (with or without etchback) used but also on the material.

SOG process integration. SOG integration for planarization above the metal layers has essentially followed two major paths, and each has demonstrated reasonable success in achieving the necessary degree of planarization. These approaches are described below.

SOG sandwich. In this technique [Yen and Rao 1988, Ito et al. 1990] the SOG is not etched back at all and is sandwiched between two layers of CVD oxide as shown in Figure 12.14. In the sandwich structure the three layers shown have specific purposes that need to be carefully evaluated when using the SOG sandwich approach.

The first insulating layer from the bottom can be either CVD oxide or oxynitride, and it serves as an adhesion and a metal hillock suppression layer that prevents the SOG from coming in contact with the metal. The thickness requirements of this first layer depend on the device design rules. As a rule

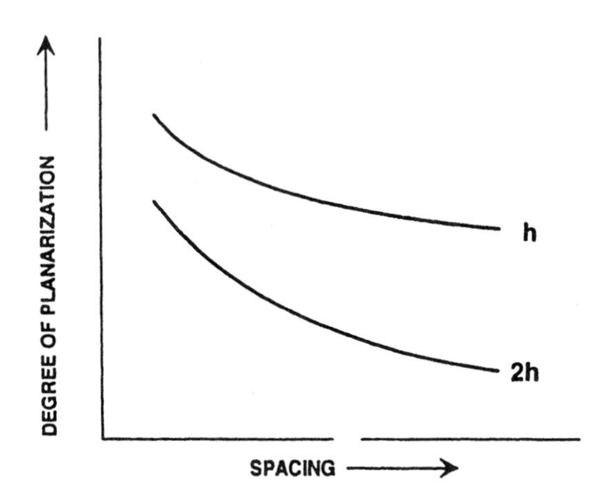

Figure 12-12 Impact of step spacing and height on degree of planarization

of thumb, the first layer should at least be able to cover the metal steps and any hillocks that may have been formed. The maximum thickness should be approximately less than half the metal spacing to prevent any formation of voids. The middle layer, whose main purpose is to fill gaps, is the SOG. In order to achieve a high degree of planarization, the SOG material selection, processing, and curing conditions must be carefully optimized. The third layer of the dielectric is the final insulating layer, and its minimum thickness is defined by the design rules for isolation and speed for the devices.

If the SOG processing and curing are not optimized, the outgassing of moisture from the SOG through the via side walls may hamper the metal deposition and thus give rise to "poisoned vias." Poisoned vias cause high via resistance or even unreliable vias. The etchback approach ensures that the SOG is removed from all potential via locations so that no poisoned vias can occur. However, this is at the expense of additional process complexity.

SOG etchback. This technique [Gupta 1989, Kawai et al. 1988, Parekh et al. 1987] employs either a partial or complete etchback of SOG after it is deposited over the wafers. The methodology is similar to the resist etchback

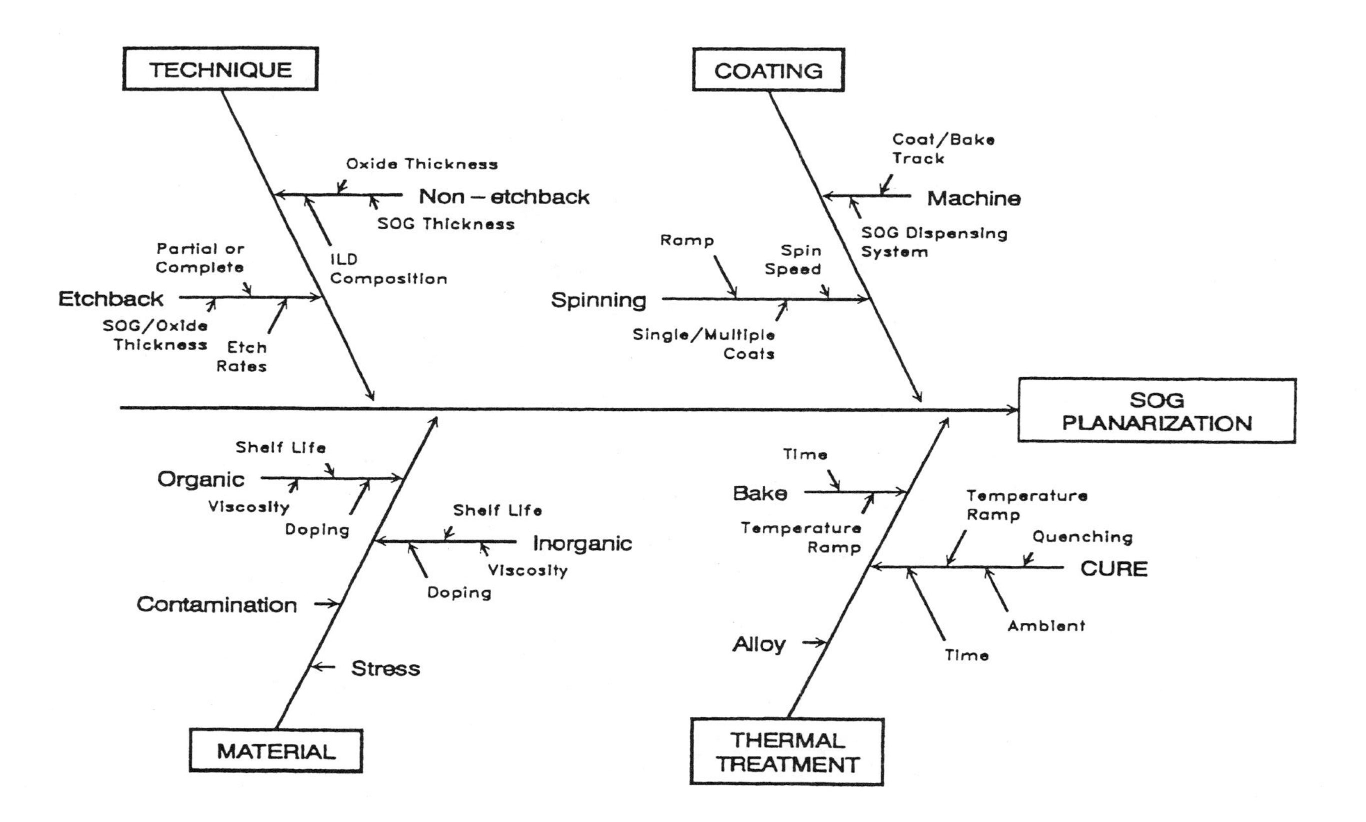

Figure 12-13 Cause-and-effect diagram for SOG planarization

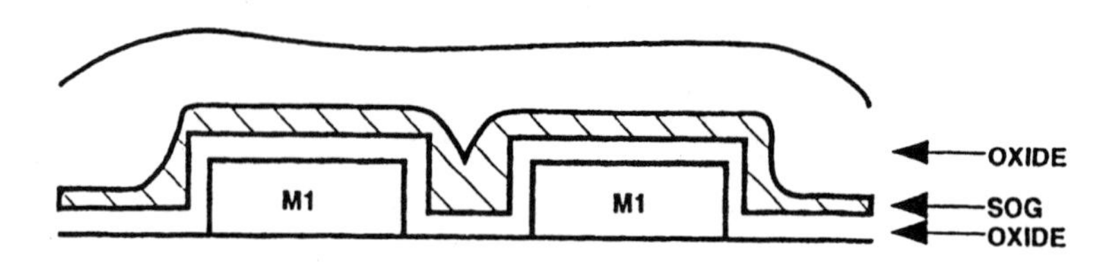

Figure 12-14 SOG sandwich structure

process. In the partial etchback variation of this technique, SOG is spun on the wafers after the first dielectric layer is deposited. The spin parameters are chosen as in the previous case to fill the space between the metal lines without any cracking. The SOG is partially etched back, targeting an etch time that leaves some oxide on top of the metal, therefore preventing the metal from being exposed during the etchback process.

During the etchback step, a selectivity of 1:1 is used to etch both the oxide and the SOG at the same rate. A desirable selectivity can be obtained by adjusting the refractive index of the dielectric layer underneath the SOG and also by varying the etch parameters, e.g., the pressure. In case of complete etchback, the SOG is completely etched away. This calls for a thicker base dielectric layer; during the etchback, the oxide and the SOG are etched back at the same rate as long as loading effects are not present. Loading effect causes degradation of planarity due to an increase in the SOG etch rate as SOG surface decreases. After the SOG etchback, a layer of oxide is again deposited in order to meet the minimum thickness requirements for the insulating film. One of the concerns of this technique is the clearing of SOG in worst-case areas, specifically on top of metal 1 where the polysilicon and the field isolation steps coincide and have minimum space restrictions.

In both the etchback and the sandwich approaches, the key to success is the manipulation of the SOG process parameters to obtain an acceptable planarity. The planarization of SOG may be affected by the planarization of the layer underneath metal 1, which in most cases is a BPSG layer. The flow angle of BPSG after reflow influences the etchback process window. To obtain a larger process window, some device technologies may demand a higher planarity of the BPSG layers than offered just by the reflow process. Postpolysilicon planarization using SOG has been demonstrated to be a viable alternative to the standard reflow processes.

Material selection. SOG is available in two different forms: silicates (inorganic) and siloxanes (organic). The chemical behavior of these glasses influences the quality of planarization. Therefore, depending on the SOG planarization requirements, the appropriate material needs to be selected. These SOGs can be doped with B or P to enhance the flow properties of the glass.

The management of the carbon content of SOG is very important for achieving good planarization performance. Increasing the carbon content decreases the film polymerization as the polymerization sites are blocked by methyl and phenyl groups. Inorganic SOG does not have any methyl or phenyl groups present and so the polymerization sites are not blocked by carbon. Increasing the carbon content decreases film shrinkage, because the polymerization sites are not blocked by carbon. Increasing the carbon content decreases film shrinkage, due to the polymerization sites being blocked. As a result, the stress in high-carbon-containing films is lower. It is thought that the crack resistance and gap filling of organic glass films is better than the inorganic because of higher surface mobility of atoms within the polymeric structures. There are fewer O-Si-O crosslinking sites and hence the film is better adapted to withstand a higher level of stress. Addition of carbon to the SOG makes the film more porous and therefore more susceptible to chemical attack, as there could be some unreacted silanol groups that react with water.

Solvents. One of the most common problems is the wettability of the SOG. This depends on the hydrophobicity of the solvent, which contains 1-propanol and ethanol. Others may also contain 1-butanol and acetone, which increase the solvent's hydrophobicity. It is believed that the SOG fills trenches primarily due to the capillary action. In some cases, depending on the solvent type, the capillary action may be counterbalanced or even outweighed by the hydrophobic dewetting forces. To enhance the wettability, the appropriate solvent type may be chosen or with the spin time can be increased, but this will be at the expense of planarity and filling.

Aging. SOGs generally have a fixed shelf-life and need to be stored in cool environments. Age causes the SOG to gel and it becomes very nonhomogenous. The usual problems in using SOG that has exceeded its normal shelf-life are striations after coating the wafers and poor film uniformity. Table 12-6 compares the properties of organic and inorganic SOGs.

Coating. The coating process depends on the viscosity and composition of the SOG and the spin speed. The speed can be uniform or ramped from a lower to a higher speed. As the SOG is spun faster the film thickness on flat wafers decreases. On topography wafers a degradation in planarity is observed, especially in narrow metal spaces with increased spin speeds. However, on top of metal lines in the worst case locations, the thickness of the SOG decreases with an increasing spin speed. On the other hand, coating the SOG with a slower spin speed affects the same parameters in

Table 12-6 Comparison of Organic and Inorganic SOGs

Property	Organic SOG	Inorganic SOG
Step coverage	Good	Fair
Passivation	Fair	Good
Crack resistance	Good	Fair
Reliability	Fair	Good
O_2 plasma resistance	Fair	Good
Application	Etchback	Nonetchback
Manufacturability	Fair	Fair

reverse. Figure 12.15 shows the effect on the planarization by varying the SOG thickness. This is achieved by spin coating several times. One method of coating SOG wafers is to start with a static dispense followed by ramp and spread cycles. This optimization is needed to ensure that a uniform film of SOG is coated on the wafers. In order to achieve the desired thickness and gap filling, double or triple spin may be required. One of the drawbacks of this procedure is that defects in the film may be generated due to nonuniform wetting of the first layer that might have been cured. In spite of this risk, multiple bakes between spins help in driving solvents out of the SOG. This eliminates the need to bake all the layers of SOG from after multiple spins as the thermal treatment may not be enough to completely remove any moisture and solvents.

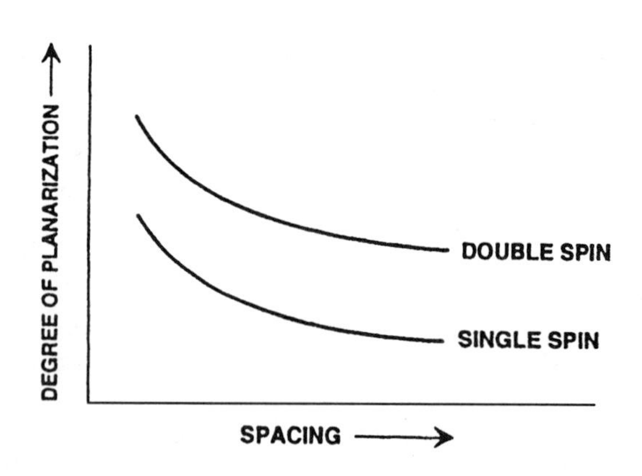

Figure 12-15 Impact of multiple SOG coatings on degree of planarization

Baking and curing. In general, two reactions [Gupta 1988] take place when the SOG is heated and cured. The first is a hydrolysis reaction where the ethanol is released from the Si-OC_2H_5 bonds. The second reaction is a condensation reaction. The thermal treatments the SOG film is exposed to drive the polymerization reaction involving the condensation of the silanol groups, resulting in the formation of the Si-O-Si bonds. During the thermal process, the solvents and water in the thermal treatment (baking after spin) cause considerable volume shrinkage in the SOG, which creates a high tensile stress in the film. This stress could contribute to crack formation in the glass. Inorganic silicate SOG is typically fragile and cracks or forms gaps in severe topography. Organic SOGs behave differently as they have methyl (CH_3) or phenyl (C_6H_5) bound to some of the silicon atoms. Presence of the methyl of phenyl groups blocks the polymerization reaction resulting in a more porous and ductile film. Organic SOGs have a higher crack resistance than inorganic ones. In order to provide sufficient time for outgassing of the solvents and moisture, the postspin bakes and cures must be performed without causing any thermal shock. Baking and curing very quickly could cause the upper layers of the spin on glass to be completely polymerized and prevent adequate moisture and/or solvent evaporation from the bulk of the SOG. Out-gassing can lead to trapping of hot volatile compounds beneath a cap of fully polymerized SOG at the upper layer of the SOG. These volatile compounds may then contribute to cracking and other forms of catastrophic stress relief. Therefore, thermal shock at either SOG bake or cure must be reduced. One possible solution is a ramped baking and curing cycle where a gradual increase in temperature occurs. By monitoring the band intensity of the O-H stretch of the SOG using Fourier transform infrared spectroscopy (FTIR) [Woo et al. 1990], it is possible to quantify the extent to which this condensation reaction approaches completion.

Control of the thermal treatment is very important in general for SOG processing and specifically for a nonetchback option because the SOG is exposed after via etch. Insufficient curing will cause moisture and solvent outgassing through the vias during the metal deposition, which gives rise to a typical defect called "poisoned vias." In such vias, the metal does not completely fill the voids and therefore causes high via resistance and reliability degradation. This suggests that any post-SOG cure temperature steps must be no greater than the SOG cure itself.

12.6 Contact/Via Formation

The objective of this process step is to open hundreds of thousands of contacts or vias on a die in order to provide a connection path between the metal interconnects and the transistors. In the processing environment, the contacts/vias need to be opened on varying topographies, across the die, within a wafer, and on several wafers in one operational step. Contact process optimization and control are being scaled down and the number of contacts are increasing per unit area. Contact placement, size, metal overlap, and other design rule constraints make the management and control of the contact/via formation step very challenging. The previous steps presented a macroscopic set of issues. Localized variations among contacts need to be studied, as these variations differ from contacts/vias that may be separated from each other by only a few microns. The challenge is to ensure consistency and control in the definition of contacts and vias using enhanced lithography, etching, cleaning, and contact filling techniques.

In Section 12.1, a brief explanation of the difference between contacts and the vias was given. As most of the issues for contacts and vias are similar, the discussion here will focus on contacts. Wherever appropriate, the discussion will also include via-specific issues. In Section 12.6.1, issues pertaining to the contact formation are discussed. In Section 12.6.2, the factors that contribute to contact resistance are examined.

12.6.1 Contact/via etch process

Contact and via etch processes [Ghandhi 1983, Ephrath and Mathad 1988, Coburn 1982] are critical steps in any integrated circuit process flow. To understand the gamut of variables that affect and modulate the contact/via etch process, a cause and effect diagram, Figure 12.16, is presented that looks at the contact/via formation process as a whole. The contact formation cause and effect diagram highlights some of the salient issues associated with the definition of contacts.

The choice of etchers, their configuration, chemistry, and machine setting determine the quality of the etch process and the throughput time of the etcher. Each of the input/machine parameters, such as chemistry, flow rate, pressure, etc. impact the various output parameters. Output parameters are those that indicate the quality and capability of an etch process. These parameters are etch rate (ER), selectivity (SEL), loading effect (LE), critical dimension loss (CD), profile (PF), uniformity (UNF), defect density (DD), and process-induced charging (PC). Let us examine each of the major machine parameters and examine their impact on the output parameters.

- *Chemistry.* The gas species used for plasma etching. Has a strong effect on the etch rate, selectivity, profiles, and CD. For etching oxides, fluorine (F)-based chemistry is widely used.
- *Mixture.* The gas ratio determines the polymer formation. For oxide etching, the fluorine to carbon (F/C) ratio is an important parameter.
- *Flow rate.* The flow rate determines the availability of the reactive species. The most prominent impact is on the loading effect.
- *Pressure.* The pressure of all the gases in the reaction chamber has a major influence on the profiles of the oxide and the critical dimensions.
- *Frequency.* Generally the frequency of the RF used is 13.56 MHz. This has an overall effect on the process, but does not modulate any single or

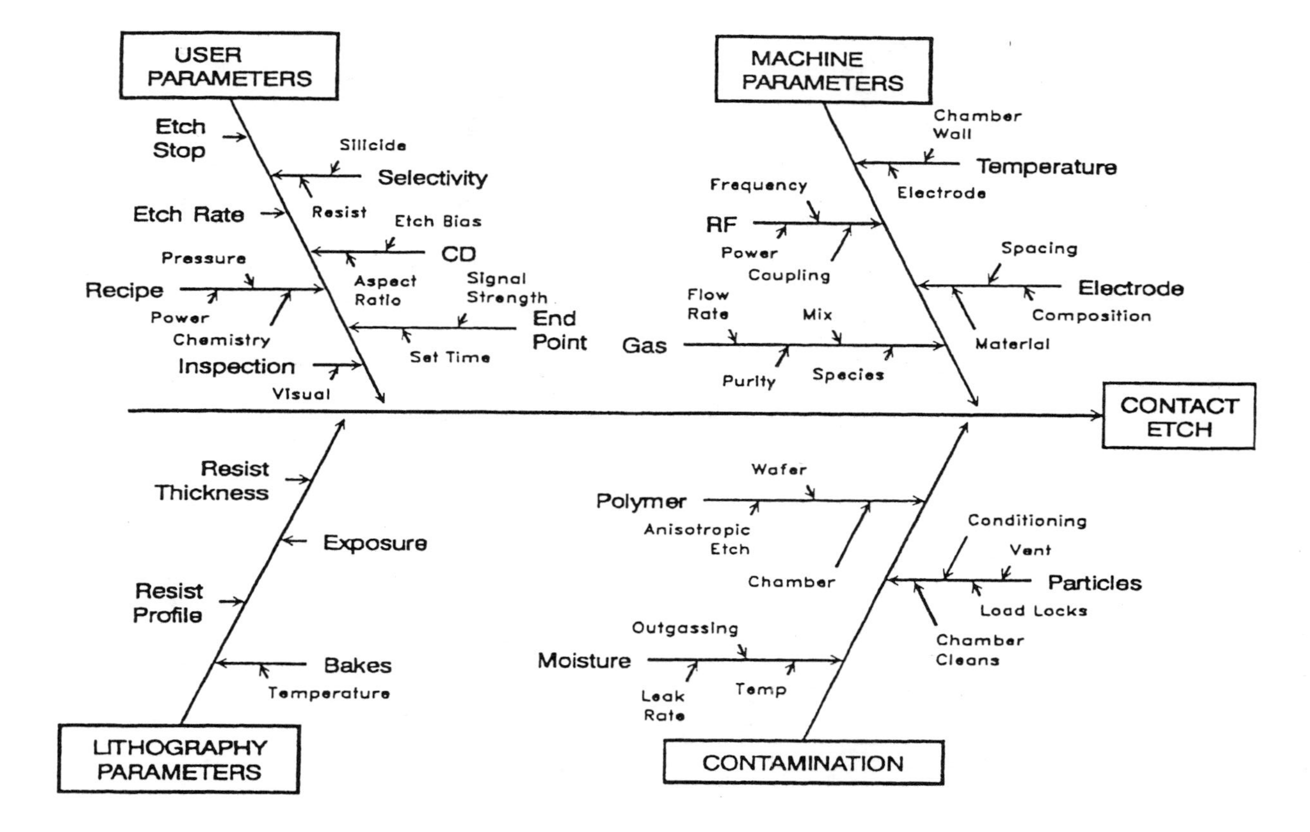

Figure 12-16 Cause-and-effect diagram for contact etch process

multiple parameters strongly.

- *Power.* Variation in power has a marked effect on the etch rates and process-induced charging.
- *Electrode.* The material, type, and spacing of the electrode primarily influences the uniformity of the etch process across the wafer.

Table 12-7 compares the impact of various machine parameters on the output parameters. H signifies high correlation and M represents medium correlation between input and output parameters. Blank spaces signify weak or no correlation.

Profile control. Tapered contacts and vias, to the extent permitted by the process sensitivity, have been used for 1 μm or larger process technologies to improve metal step coverage. In submicron technologies, tapered contacts are not a viable option as contact critical dimensions, spacing, metal overlap of contacts, and other design tolerances have been tightened due to device scaling.

How can we achieve near-vertical contact/via profiles? Near vertical contact/via profile control is generally achieved by defining near vertical resist profiles and using a RIE process. In addition, managing polymer buildup during the etch process, through appropriate selection of the F/C ratio [Coburn and Winters 1979], also helps obtain the desired sidewall angles. The polymer coats the sidewall and prevents lateral etch. Removal of the polymer after etch may pose a problem. In the case of contacts, acid cleaning may be used to strip the polymer, but for vias the choice of cleaning agents decreases as acid cleaning would attack the metal layers underneath. Oxygen is added to the fluorocarbon gas in order to improve the etch uniformity, enhance etch rate, and reduce polymer buildup. Reduction in polymer buildup is needed for decreasing the particle density on the wafers and also for minimizing the frequency of the etch chamber cleaning. The profiles of the contacts and vias determine the quality and integrity of the contact/via filling process. For example, a blanket CVD W fill deposition process is very sensitive to the contact/via profile and one can expect degraded W step coverage along with a void in bowed or reentrant contacts/vias.

Table 12-7 Correlation between Machine and Output Etch Parameters

Parameter	Etch rate	SEL	LE	UNF	CD	Profile	DD	Charging
Chemistry	H	H	M	M	H	H	M	
F/C ratio	H	H	M	M	H	H	M	
Pressure	M	H		H	H	H		M
Frequency		M		M	M	M		M
Power	H	H		M	H	M		H
Electrode	M			H	M	M	M	
Flow rate	M	H	H		M	M	M	

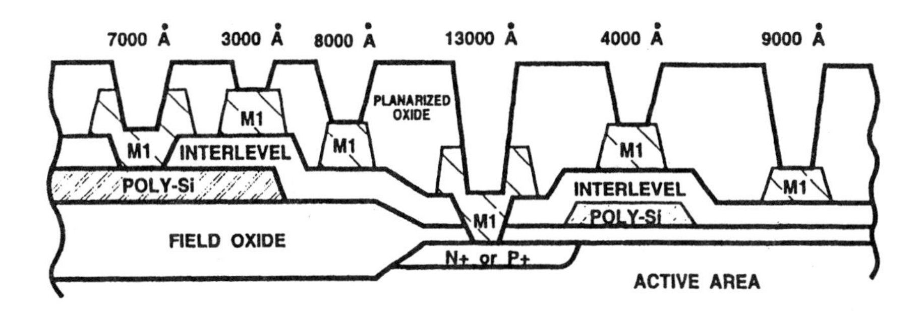

Figure 12-17 Via depth variation in a multilevel interconnect system [Brown et al. 1986] (© *1989 IEEE.)*

Differential depths. The planarization section discussed the advantages of planarization. A side effect of global planarization is the creation of contacts and vias with varying depths. Differential depths present a unique contact/via etch scenario. Figure 12.17 shows the via depth differential and the scenario after etch that is a consequence of the planarization [Brown et al. 1986]. Overetching of the shallow vias cannot be prevented as the deep vias take a longer time to be etched. New interconnect materials may be the answer to this problem, along with an effective contact and via cleaning processes. Then cleaning processes are done prior to metal deposition. During this process, both wet and dry cleaning processes are used.

Contact etch stops. As the implanted source/drain and gate regions offered an unacceptable contact resistance for submicron devices, the use of silicides such as $TiSi_2$ and $CoSi_2$ became necessary. In the contact etch process, for example, the etch must stop on the silicide layer. To prevent significant etch of the silicide in the contacts, especially in the shallow contacts, the selectivity of the silicide film must be very high [Jaso et al. 1989]. If the shallow contacts are overetched so that the salicide layer is also etched, the contact resistance will increase drastically. It has also been observed [Jaso et al. 1989] that addition of H_2 to the fluorocarbon gas chemistry greatly enhances the selectivity of the $TiSi_2$ film without markedly affecting the oxide etch rate. $SiO_2/TiSi_2$ etch rate ratios of 40:1 have been reported by the formation of a fluorocarbon film over the silicide film. The margin for process control may be small as contact and via etches of effective endpoint systems are still needed. Using endpoint systems for these etches is complicated by the fact that contact/via areas are small.

Via etch stops. The vias contain a totally different etch stop. The underlying layer that acts as an etch stop is a metal stack. If aluminum is the stopping layer then, it reacts with the fluorocarbon gases to form a Al-F compound. In addition, other aluminum materials may be formed in the vias during the subsequent contact metallization steps that may also increase the via resistance.

12.6.2 Contact resistance

Contact formation and its subsequent filling with metal (discussed in Section 12.7, below) significantly modulate the contact resistance. This is monitored either inline or usually at the completion of the process on the screens to check the functionality of the finished device. The contact resistance measurement is done on specially designed test structures, containing hundreds of contacts in a chain, and placed in the silicon streets between the dice on the wafer. The variations in contact resistance are used as a flag to diagnose any problems with the contact formation or the contact filling process steps. Control of this parameter is one of the ways to monitor lot-to-lot performance of the process in general. Notice that there are several other monitors in a process that check the performance of the various steps. The contact resistance test is an integrated monitor and any out-of-control situations would necessitate a closer examination of individual step monitors, at least during the initial stages of failure analysis. To put the whole contact resistance issue in perspective, a cause and effect diagram is given in Figure 12.18.

In an ideal situation, what contributes to contact resistance or via resistance? In the case of a contact, it is mainly the sum of the interfacial (interface between contact metal and silicide) resistance and the silicide resistance. In the case of a via, it is primarily the interfacial resistance. Assume the entire contact (ideal in formation and with unit dimensions) right from the metal to the silicide layer is a block, with each section having its own resistance, as shown in Figure 12.19. The overall contact resistance is the sum of the interfacial resistance (R_i) and the silicide resistance (R_s), if we assume that the contribution of the plug (R_w) and the metal line (R_m) is negligible. In other words, total contact resistance, R_c, is given by

$$R_c \approx R_i + R_s \qquad (12.12)$$

R_s depends on the silicide formation process and is well controlled. R_i is basically a measure of the contact process health as the significant portion of the contact resistance comes from the interface of the W plug and the silicide. The interface is altered by the etch process, sputter etch, and the W adhesion layer interactions, as explained in Section 12.7.

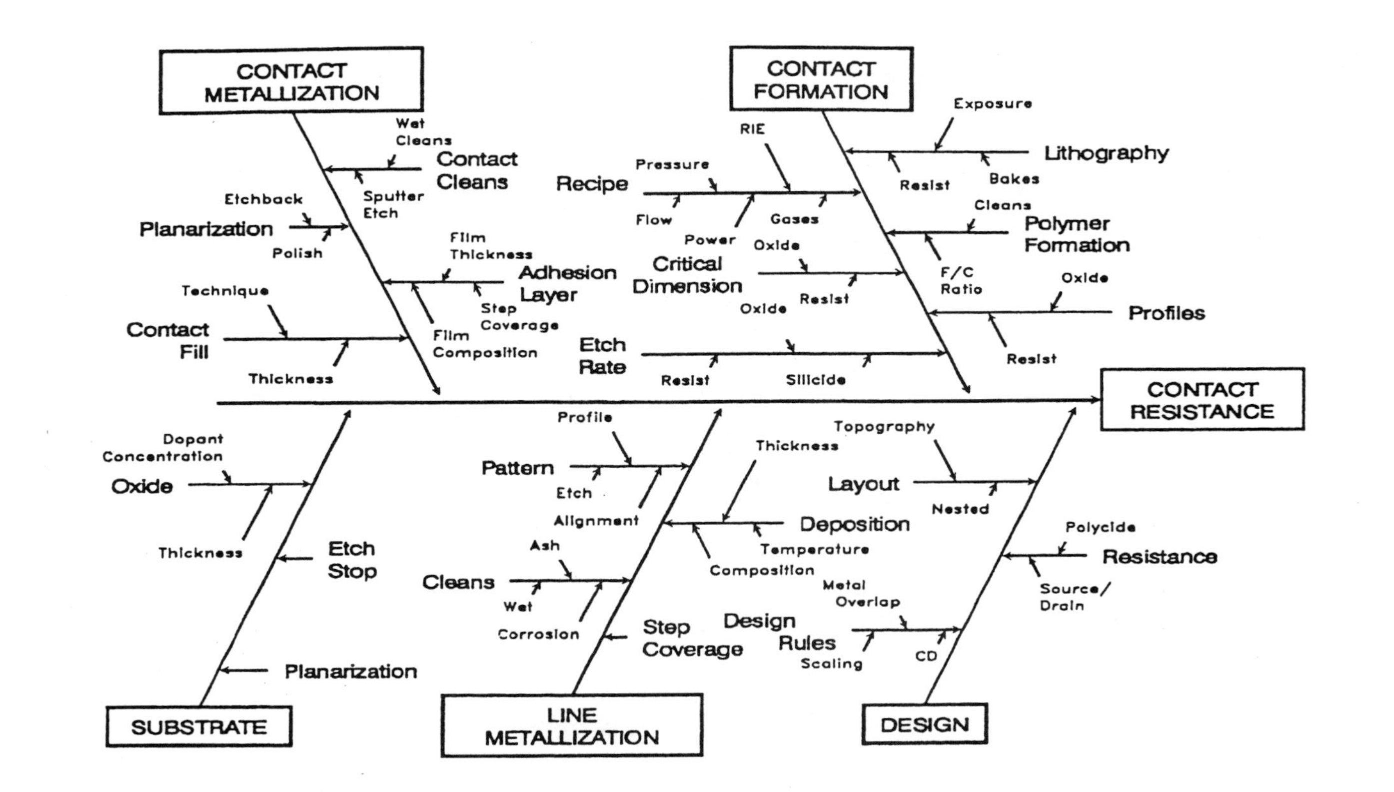

Figure 12-18 Cause-and-effect diagram for contact resistance

LINE METALLIZATION RESISTANCE	R_m
W PLUG RESISTANCE	R_w
INTERFACE RESISTANCE	R_i
SILICIDE RESISTANCE	R_s

Figure 12-19 Contribution to contact resistance by the components in a tungsten-filled contact

We can use some generic illustrations to understand the relationship between the contact resistance (applicable to via resistance also) and the following parameters. This list of relationships is not comprehensive but is intended for the reader to appreciate the interactions.

Aspect ratio. Figure 12.20 shows the relationship between contact resistance and the aspect ratio. The *aspect ratio* of a contact is defined as the ratio between its depth and width. If the aspect ratio of a contact is high, it poses a problem for contact filling. Contact resistance increase may then be dominated by the integrity step coverage of the contact and line metallization.

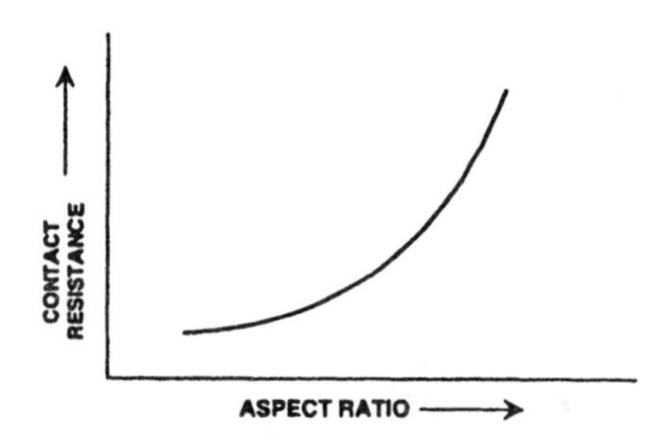

Figure 12-20 Impact of aspect ratio on contact resistance

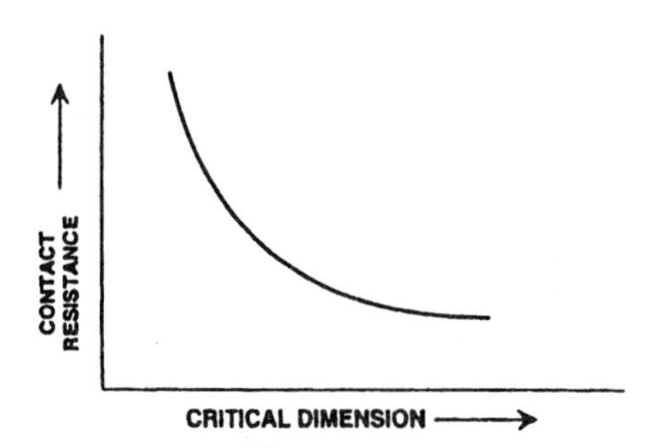

Figure 12-21 Impact of critical dimension on contact resistance

Critical dimension. The contact size, the cross-sectional area at the bottom of the contact, influences the amount of current flow. As the critical dimensions (CD) of the contacts decrease, this cross-sectional area also decreases, offering a higher resistance to the current flow. As in Figure 12.20, this relationship is valid as long as the contacts can be defined in resist, etched, and filled with metal without any loss of contact integrity. The interaction between contact resistance and the critical dimension is shown in Figure 12.21.

Overetch. Since contacts have to be etched with differential depths, overetching is essential to ensure that all the contacts are opened and therefore the contact resistance decreases. Figure 12.22 shows the contact

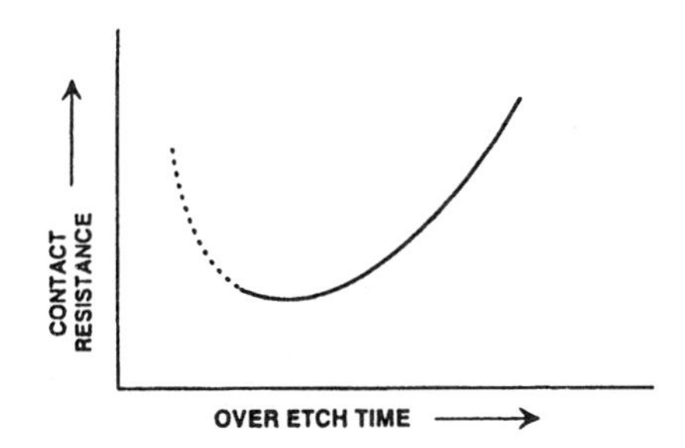

Figure 12-22 Impact of contact etch time on contact resistance

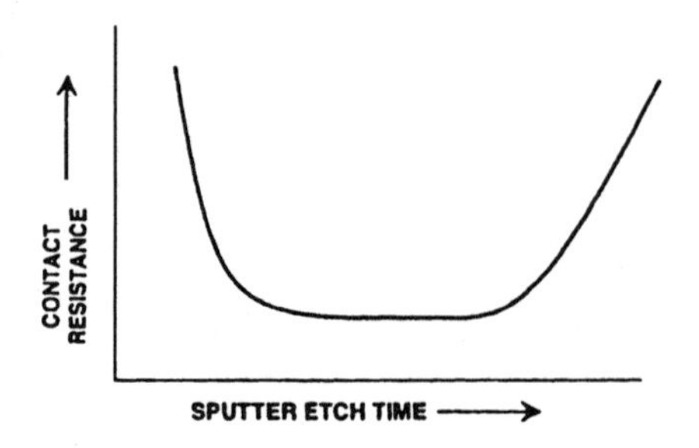

Figure 12-23 Impact of sputter etch time on contact resistance

resistance as a function of the contact overetch time. However, overetch also causes the shallower contacts to be exposed for a much longer time to the etch. This does not necessarily create problems as long as the selectivity of the underlying layer is high with respect to the main film that is being etched. As the overetch time increases, the contact resistance also increases and the bottom of the contact is damaged and the etch stop starts to be etched more and more. This relationship is illustrated by the bold curve in the graph in Figure 12.22.

Sputter etch time. Just prior to contact metallization, any native oxide is removed by using argon (Ar) sputter etch. The sputter etch process is a physical etch process employing accelerated AR ions. As the sputter etch time is increased, the contact resistance decreases up to a point. Beyond this, any further increase in sputter etch time merely causes the redeposition of material from the sidewalls and increases the defect density. Consequently, the contact resistance starts to increase. Figure 12.23 illustrates the relationship between contact resistance and contact sputter etch time.

This section has explored the issues with the contact formation process and what modulates contact resistance, which is an important parameter for determining the state of the process. The next section will discuss contact metallization and interconnect metallization. How these two influence the contact resistance has been shown in the contact resistance cause-and-effect diagram in Figure 12.16.

12.7 Metals

This last section in the multilevel interconnect process flow (Figure 12.2) deals with contact and interconnect metallization. In terms of process technology development, contact metallization is a recent development that was motivated by the migration of device geometries toward submicron dimensions. As this trend continues, more and more development effort will be devoted to understanding the materials and their metallurgical interactions with silicide or other underlying layers in the confines of a contact or via. In fact, contact metallization has two basic components–contact filling and contact planarization–and each is a subject for extensive research and development. Line or interconnect metallization has been studied for a long time. A wide variety of conductor materials are available for metal interconnects that use aluminum (AlSiCu) alloys and also sandwich structures using refractory and other metals, which are examined in Section 12.7.4. For submicron process technologies, the key issue under consideration is the current-carrying capability of the interconnects and other process-induced defects. The discussion below is divided into two portions, contact and line metallization.

12.7.1 Contact metallization

The design rules for submicron process technologies are rapidly changing the vertical contacts and vias that are being presented for metal deposition. In previous technologies, contact and via filling with good metal step coverage was not an issue as these were tapered. As explained above, tapered profiles of the contact/vias places limits on the maximum packing density. Profile control is difficult to achieve in a manufacturing environment. Vertical contacts and vias cause metal step coverage to be poor if the metal was deposited directly into these holes. In order to improve the step coverage of the metal interconnects, the contact/via holes have to be filled with material that can fill the holes without any voids and can be planarized [Lee et al. 1989, Jucha and Davis 1988, Blewer 1986, Wu et al. 1989]. Metal interconnects running over these plugs will have a superior step coverage and will also be able to conduct current through the contact or via.

12.7.2 Techniques

There are many possible approaches to contact metallization. The most attractive technologies are the blanket tungsten deposition/etchback and selective tungsten deposition. There are still a lot of manufacturing issues that need to be resolved with these processes. However, until the other approaches are more manufacturable, these two contact metallization methods may continue to be used for the fabrication of submicron devices. The approaches [Kotani 1988] for contact metallization may be categorized as follows.

Blanket CVD. Thick conformal deposition of metal can be accomplished by using low-pressure CVD techniques. The thick film may be etched back or polished just for contact filling or may be patterned and the interconnect metallization defined. Such a deposition fills holes and grooves conformally, and the deposition rate is limited by surface reaction and not by mass transport. The main contenders for this method are W and CVD-Al. Although both processes are still in a continuous mode of development, blanket W has gained wide acceptance because of its manufacturing ease.

Bias sputter. Al and molybdenum (Mo) films have been used to fill contact holes by bias sputtering under high resputtering conditions. One of the drawbacks of bias sputtering is the extremely low deposition rate; hence it is not a viable option for a high volume manufacturing process. Barrier layers are necessary to prevent Al and Si interaction. High deposition rates, improved film quality, and contact hole filling over a wide range of aspect ratios would make bias sputter deposition of metals (Al, Mo) a viable alternative to other contact filling techniques.

Selective deposition. Selective deposition has an unique property. The film deposition takes place over SI, silicide, and other metals but not over dielectric materials. Selective W deposition has some obvious advantages as explained in greater detail below. However, one of the problems of selective W deposition is filling contacts with varying depths and substrates.

Other. Other approaches that have been proposed are the pillar formation, lift-off, blanket CVD polysilicon, and electroless metal deposition. Even though some of these may appear elegant, their use is determined by their manufacturability ease.

12.7.3 Contact fill and planarization

The two approaches that will be discussed in some depth are blanket W deposition with etchback and selective W deposition.

Blanket tungsten deposition and etchback. As blanket W is widely used to fill submicron contacts and vias in manufacturing processes, let us review the cause and effect diagram for contact filling and planarization using blanket W deposition and etchback as shown in Figure 12.24. W does not adhere to dielectric substrates; therefore an adhesion or glue layer is needed for the W to nucleate on [Wolf 1990]. There are several variations to the adhesion layer. A couple of the commonly used adhesion layers are TiN and TiW. It is this adhesion layer that comes in contact with the silicide source and drain region and contributes to the interfacial contact resistance. The step coverage of the adhesion layer in high-aspect ratio contacts and vias is of concern as poor step coverage may cause W-related defects if it reacts with underlying layers. This issue may be overcome by combining conventional glue layers with CVD-WSi_x or other thin films.

In the deposition process, [Yu et al. 1990] W is deposited in a low-pressure, low-temperature CVD system using tungsten hexafluoride (WF_6). The chemical reaction involves the reduction of WF_6 using both H_2 and SiH_4 in a two-step process to form W. The deposition process has to be carried out in two steps employing both the properties of silane and hydrogen reduction chemistries. The majority of the film is deposited by the hydrogen reduction of WF_6 after a few hundred angstroms are first deposited by silane chemistry. The two chemical reactions proceed as follows for the silane and hydrogen, respectively:

$$2WF_6 + 3SiH_4 \rightarrow 2W + 3SiF_4 + 6H_2 \qquad (12.13)$$

$$WF_6 + 3H_2 \rightarrow W + 6HF \qquad (12.14)$$

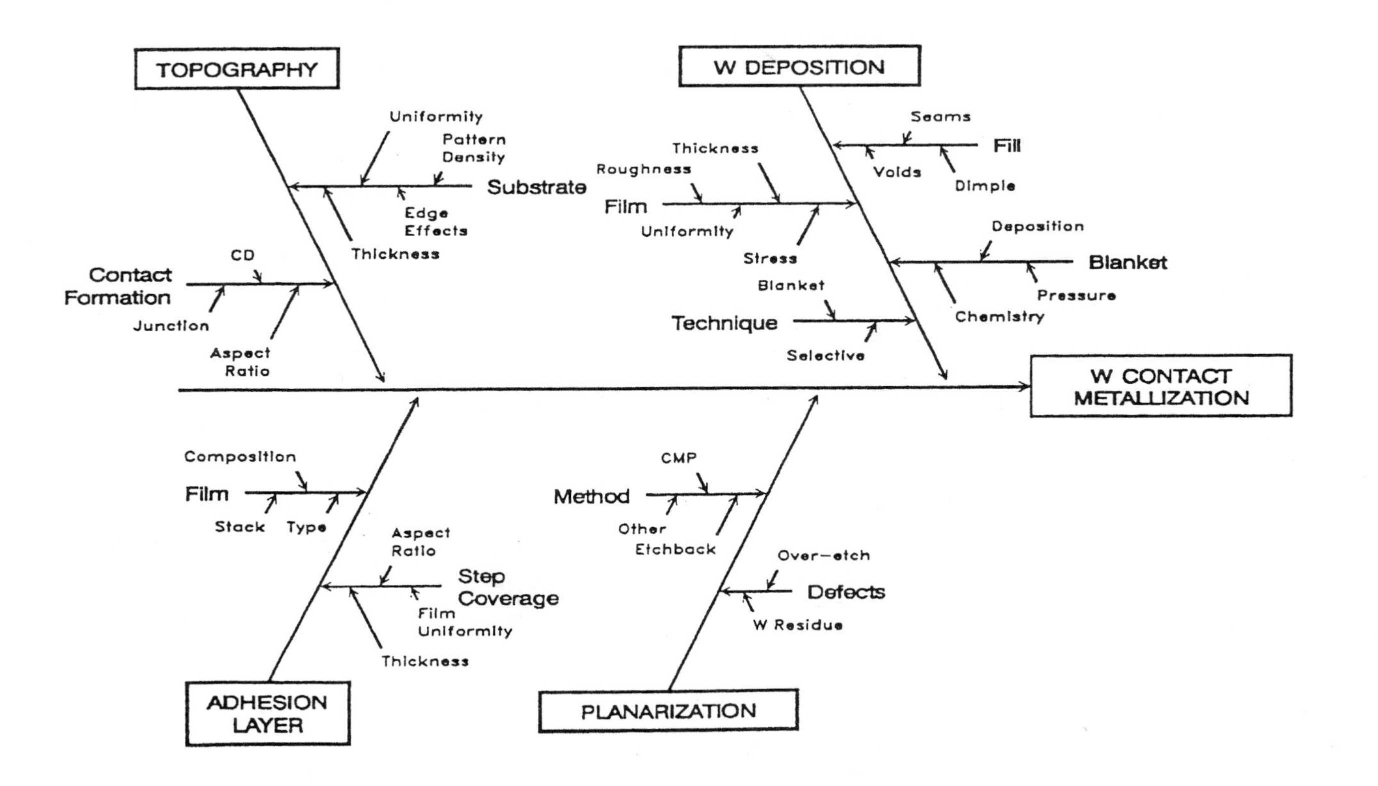

Figure 12-24 Cause and effect for contact planarization using tungsten

Table 12-8 Comparison of W Deposition Processes

Parameter	Silane	Hydrogen
Deposition rate	High	Low
Step coverage	Poor	Good
Nucleation	Glue layer	W
Thickness	Very thin	Thick

Table 12-8 puts the benefits of each of these reactions in perspective. The first step in the blanket deposition process is the deposition of a very thin W film. This film is a few hundred angstroms thick and the reaction proceeds using silane, as shown in Equation (12.13). After this step, the bulk of the W film is deposited by hydrogen reduction of WF_6. The W must be deposited without voids in the contacts.

One of the drawbacks of the blanket deposition is the formation of dimples in the center of the plugs. This dimple formation is not only dependent on the deposition process, but also on the contact geometry. The W film is rough. Roughness of the film impacts the dimple formation in the W plugs. The rougher films seem to have a smaller dimple size as compared to smoother films.

In the case of via filling, blanket tungsten is again used. However, if the tungsten via plugs are formed on Al interconnects, the control of via resistance is more difficult than if the plugs were to be defined over W interconnects. During the adhesion layer deposition and the subsequent W deposition process, nonvolatile intermediate species, e.g., aluminum halides, form at the bottom of the via on the Al surface. The presence of these compounds may increase the via resistance. Figure 12.25 shows a scanning electron micrograph of the W film after deposition in a contact.

As the W is deposited all over the wafer, it must be etched back, leaving W only in contact and via holes. Several techniques [Wolf 1990] have been demonstrated to perform this planarization, like chemical mechanical polish [Guthrie et al. 1991]. Each of these has its own technical and manufacturing problems. Tungsten etchback is still widely used for contact planarization. The simplest way to etchback is to etch the W directly, without using any sacrificial layer. The process does appear to be simple in concept, but has several issues that may affect the contact/via resistance and metal step coverage causing yield degradation.

Bulk film etch. The incoming W film thickness is dependent on the smallest and the largest aspect ratio contacts. W must fill these contacts

Figure 12-25 Cross section of a tungsten film in a contact

without any voids and have sufficient thickness for the etchback process. As the contact holes become narrower, the W deposition process becomes difficult as voids may form. Wider contacts, on the other hand, need a greater amount of material to completely fill the contacts. The W etchback process is performed using fluorine- and chlorine- (for example, SF_6, Cl_2, and He) containing gases. The purpose is to leave a planar plug in the contact or via with no residue [Sivaram and Mu 1989]. Usually the etchback process is done in multiple steps. The first step is aimed at etching the bulk of the W film in an isotropic manner with a high etch rate. To achieve an isotropic etch, fluorine-based gases are employed. The isotropic nature of this step also smoothes the inherent roughness of the W film. The second step is an anisotropic etch employing chlorine-containing gases with a high selectivity to the adhesion layer. Further discussion on the adhesion layer etch is presented below in the overetching section.

As in the contact etch process, shallow contacts may be severely overetched during the etchback processes [Wolf 1990]. This may cause catastrophic failures of shallow contacts as the W and the adhesion layer get severely etched from the plug. As the adhesion layer is etched out, the underlying silicide layer is exposed to the etch chemistry and is removed. Removal of the silicide layer causes the contact to have a high resistance and therefore fail. The polishing of W would overcome such problems.

Microloading. This is a problem quite specific to the etchback process. Localized variations in etch rates can exist and cause areas of complete removal of plugs while at the same time there may be areas where the W is still not fully etched back. Microloading occurs during the final clearing of the W over the dielectric substrate. Microloading is a localized effect where the etch rate differences occur due to localized consumption or depletion of reactant neutrals and appear to be independent of the ion-assisted surface reaction processes.

Overetching. Overetching is essential to ensure all the W residues are completely removed. Residues left on the wafer may cause interlevel shorts and other defects that may nucleate on these sites. During the overetch step, the selectivity of W to the adhesion layer etchout depends not only on the selectivity but also on the thickness of the adhesion layer itself. The nonuniformity of the overall W deposition and etchback process will impact the contact/via resistance variation and even the step coverage of the metal interconnects. The schematic diagram shown in Figure 12.26 illustrates a possible scenario if the adhesion layer were to be overetched that could lead to reliability degradation [Wolf 1990, Schmitz 1992]. The etchback process may be combined with the definition of X interconnects, instead of AL. W interconnects, along with the rest of the materials that may be used as interconnects, are discussed in Section 12.7.4 below.

Selective W deposition. This method [Broadbent and Stacy 1985, Maluf et al. 1990, Singer 1990] of filling contacts and vias has been investigated by many researchers and it should become a viable process once some of the manufacturing issues are resolved. Selective deposition allows W to form on

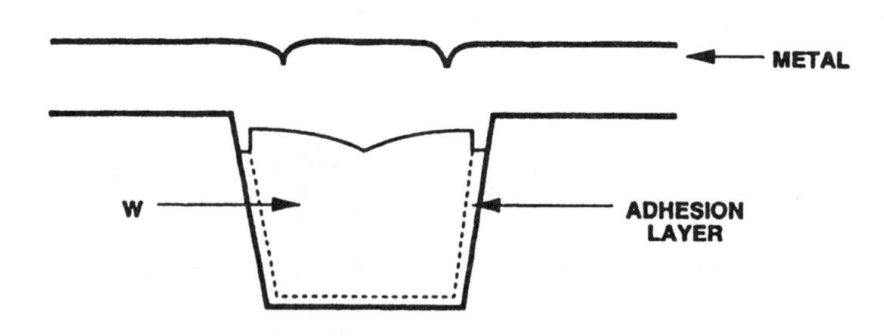

Figure 12-26 Impact on metal step coverage due to blanket tungsten deposition and etchback process variations

silicon, silicide, or metal areas in the bottom of contacts or vias and not over dielectric films. The W is formed by hydrogen reduction of WF_6, as given by Equation (12.15), only after an initial reaction of WF_6 with Si in the contacts. The chemical equation for this reaction is given as follows:

$$2WF_6 + 3Si \rightarrow 2W + 3SiF_4 \qquad (12.15)$$

Prior to the deposition of W, any silicon dioxide is removed from the bottom of the contacts and then the reaction is allowed to proceed. During the initial few minutes, the W deposition occurs by the Si reduction of WF_6, as shown above. This is a self-limiting step; after a few tens of nanometers of W are formed, the reaction is stopped as the W acts as a diffusion barrier and prevents the Si from reaction with the WF_6. After this initial phase, the remaining amount of W required is formed by the hydrogen reaction, as shown in Equation (12.15).

Improvement of selectivity is necessary to ensure that only the Si, silicide, or metal layers are deposited with W and not SiO_2. The selective feature of W deposition depends on numerous parameters. Degradation of this feature is observed when WF_6 or H_2 partial pressure, deposition time, deposition temperature, and the ratio of the Si to SiO_2 area are increased. This suggests that the management of the entire deposition is crucial to ensure that the selective property is maintained.

Besides improvement in deposition, another technique [Cheek et al. 1991] prevents W formation on the dielectric oxides. This is accomplished in two ways. First, change the surface of the oxide just prior to W deposition. The oxide surface may be altered by a HCl plasma or methanol pretreatment [Cheek et al. 1991]. This has the effect of controlling the dielectric surface states that affect the selective deposition property. Second, cover the oxide with some other material that suppresses W nucleation.

Cleaning of the contact/via surfaces in needed to improve the contact/via resistance as any residual contamination or native oxide present on these surfaces may prevent the formation of W. Recent work [Ohba et al. 1991] suggests that the use of dimethylhydrazine (DMH) gas in the CVD W deposition process can remove the native oxide of the Si surface. DMH cleaning showed a drastic reduction in the sheet resistivity of the W layer and had some positive effects on the junction leakage.

12.7.4 Line metallization

This is the last step in the sequence of operations for multilevel interconnect formation. After all the interconnect layers are defined, a passivation film is deposited and the device is tested for functionality. After testing the devices, proceed to the assembly sites for packaging. Interconnects serve the following function:

- Provide paths to contact junctions and gates
- Interconnection between device cells
- Input and output pads for connection to external signal sources via the package leads.

To satisfy these functions, interconnects should have a low resistivity for faster signal propagation, be reliable under various operating conditions, and adhere well to Si and SiO_2. Moreover, they should be compatible with the other fabrication processes.

Aluminum interconnect problems. As mentioned above, several materials are being used for the interconnects. The question is, Can we continue to use Al interconnect schemes for submicron devices? Several technologies have begun to use W and even Copper (Cu) interconnects as an alternative to aluminum because of the problems associated with Al interconnects [Arita et al. 1989, Yamamoto et al. 1987].

Aluminum silicon alloys have become the primary materials for interconnects because pure aluminum has great affinity for silicon and therefore consumes it from the junctions. To prevent this, Al has been saturated with Si. Even though the problem of junction spiking was solved with the use of an Al-Si alloy, the problem of Si precipitation still exists in contacts that do not use barrier metallization. These precipitates have a detrimental effect on the contact resistance and also on the electromigration performance of the interconnects. For enhancing the electromigration performance of the interconnects, Cu is added to Al. The amount of Cu that can effectively prevent electromigration degradation is typically around 0.5 percent. Any higher percentages of Cu cause difficulties in dry etching, corrosion, and ultimately reliability degradation. Besides electromigration issues, Al interconnects have suffered from their susceptibility to form hillocks (stress relief mechanism) and voids.

In spite of all the problems with Al interconnects, current metallization schemes have attempted to make them more robust by adding shunt layers like Ti and barrier layers like TiW or TiN. For lithography improvements, an antireflective layer is added to the stack. The current submicron processes are not just two-level or three-level metal systems anymore. In fact, on closer examination, the metal stack often consists of several metal layers. Therefore, several metal CVD or sputter machines must be used for the deposition of the metal layers starting all the way back at the silicide process step.

Cause and affect analysis. There are several interrelated process steps that can influence the quality of the interconnect definition. The cause and effect diagram, shown in Figure 12.27, puts these in perspective. It is the time delay for signals to propagate through the performance of the device and not so much the switching speeds of the transistors. Therefore, metallization becomes a very key component of VLSI/ULSI technology because it impacts the device density and performance.

Let us discuss the aluminum interconnect definition by using the cause and effect diagram that is given in Figure 12.27. This diagram may have to be modified should other materials be used for interconnects, but the basic approach will still be the same. Substrate and contact metallization issues were discussed in earlier sections with more detailed cause and effect diagrams and in-depth analysis of the issues. Therefore, let us focus on the two important items of interconnect definition, metal deposition and metal etch/cleans.

Metal deposition. The quality and integrity of metal deposition is extremely critical for ensuring high yielding and reliable parts. Another cause and effect diagram, shown in Figure 12.28, is used to explain the integrity of sputtered metal. Metal is deposited on wafers by sputtering Al and other elements from a common target material. In the sputter chamber, Ar ions (used for sputtering) dislodge the Al wafer. The appropriate target material with the desired composition is selected since it is the source material for the metal deposition. It erodes with time as sputtering progresses. Sputtering produces films with an excellent compositional uniformity across the thickness of the film.

Heating the substrate improves the step coverage because the surface mobility of the deposited metal atoms is proportional to the temperature. This helps in improving the step coverage, but only to a certain limit. The importance of step coverage in thin films has been discussed above. In the case of Al alloys, step coverage of the film is strongly dependent on the temperature the wafer is exposed to during deposition. As temperature increases, surface mobility of the deposited metal atoms increases and therefore the step coverage is enhanced. However, an increased deposition rate may degrade the step coverage because surface mobility rates of atoms on the wafer surface are decreased due to the arrival of more and more atoms. Increased heat causes hillocks to form in aluminum films due to stress relief mechanisms. Another property that is changed by temperature is the grain size of the metal. The grain size variation is not just dependent on temperature but also on the surface topography, sputter deposition conditions, and vacuum integrity.

The sputtering process is complicated since a lot of variations in the sputtering parameters may produce films of an inferior quality. For example, the sputtering process depends on the equipment used, and the equipment parameters as shown in Figure 12.28. The film properties depend on the integrity of the vacuum system used. The base pressure on any vacuum system indicates the level of system cleanliness and the leak rate in the system chamber. During the deposition of the metal, any leaks or chamber outgassing will alter the film properties. The residual gas contribution to the sputtering gas may result in the incorporation of nitrogen and oxygen in the metal films. Refractory metals, aluminum and its alloys are readily oxidized and the resulting film resistivities and stress may be high. Residual gas contamination may cause a decrease in film reflectivity

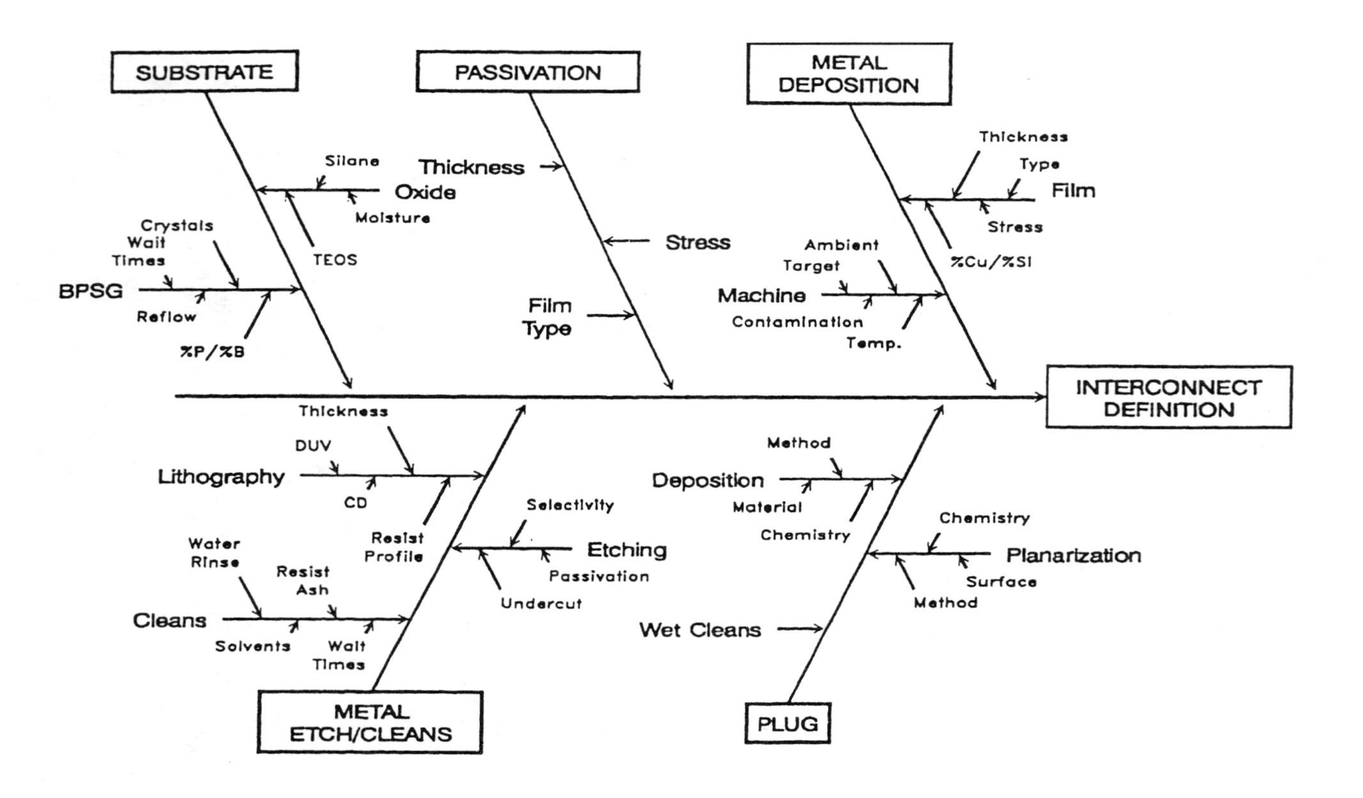

Figure 12-27 Cause-and-effect diagram for interconnect definition

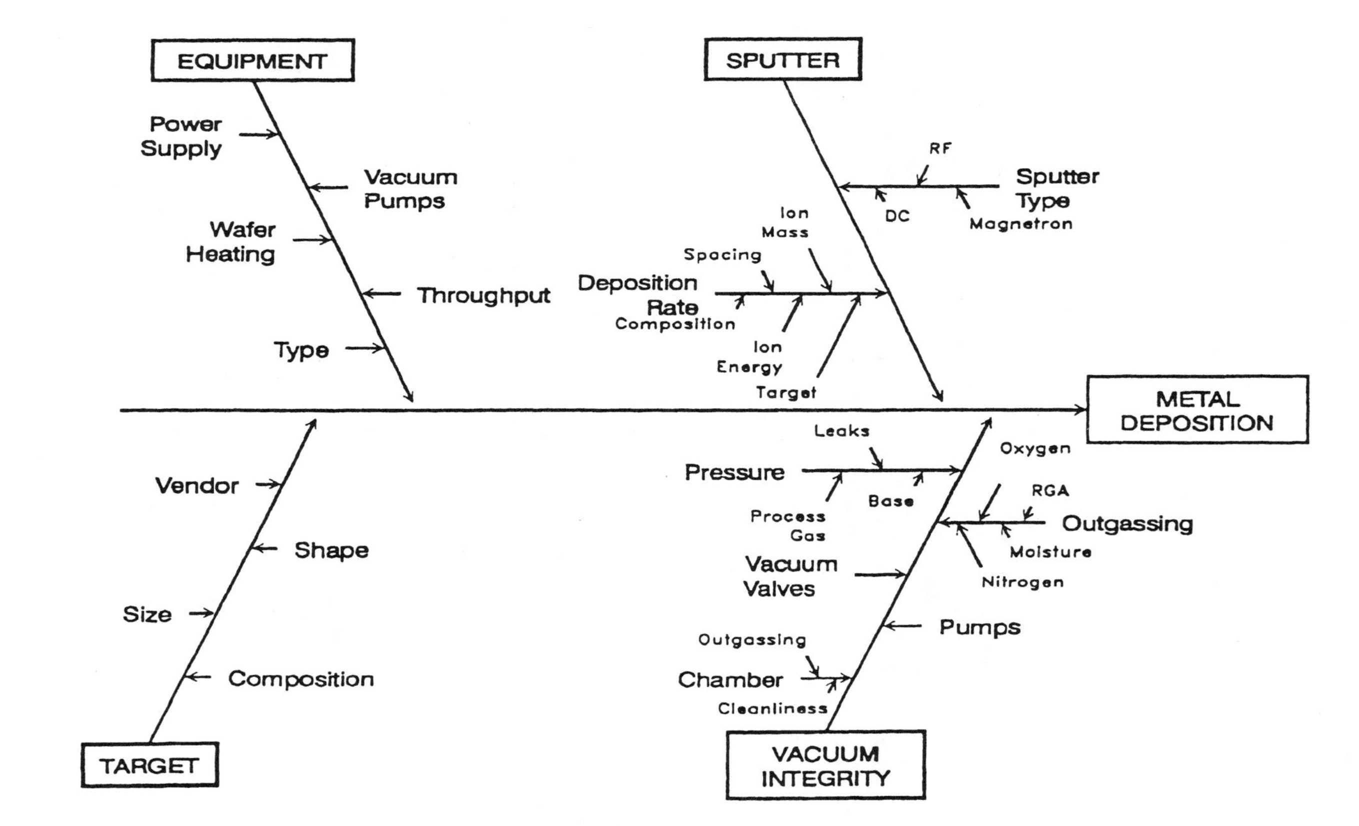

Figure 12-28 Cause-and-effect diagram for metal deposition

also. A residual gas analyzer (RGA) may be used to monitor the chamber conditions prior to deposition.

Metal etch. By far the most crucial aspect of interconnect definition is the process of etching [May and Spiers 1988, Hu et al. 1990, Lutze et al. 1990] Al alloys with shrinking pitches and tight critical dimension control. The etching of Al alloys is very important and difficult as poor management of pre-etch, etch, and post-etch treatments could cause metal corrosion and other problems, specifically in Al alloy interconnects. Corrosion is a major problem that affects metal interconnect integrity and has been a subject for extensive study [Cameron and Chambers 1989]. Hence, let us examine the issues associated with metal etch from a corrosion perspective. The whole sequence of interconnect definitions may be divided into three parts, that is, premetal etch, bulk etch, and postmetal etch.

Premetal etch. The pre-etch processing deals with the resist processing. The resist type (e.g., vendor, single, or multiple layers) impacts the etch process and the output parameters, such as line widths, etc. This becomes more of an issue with the definition of each interconnect level. The metal thickness increases depending on the interconnect level usage. With increased thickness comes a demand for increased resist thickness, tight critical dimension (with decreasing metal pitches), and profile control. All this has to be done on a highly reflective surface (even with the use of an antireflective layer) which may not be perfectly planar (depends on the planarization process). Increased metal thickness translates to increased etch time and a chance for resist profile degradation during the etch. The resist has to harden after exposure and development for it to withstand the etch process. Unhardened resist will reticulate during the etch. Hardening is accomplished in a deep ultraviolet (DUV) system where the temperature and exposure to UV light form a crust on the resist. The crust protects the bulk of the resist and prevents reticulation.

Bulk etch. To pattern Al and its alloys, chlorine (Cl)-containing plasmas are used in a RIE mode. This form of etching has been quite successful in giving good metal profiles and tight critical dimensions, both of which are key requirements for increased packing density of interconnects. But using Cl chemistry to etch aluminum is not without major problems. Al and Al alloys patterned in such a manner have been found to corrode very rapidly on exposure to the atmosphere. The etching is further complicated with the use of different materials to strengthen the interconnect capability of the aluminum, in the form of barrier, shunt, and antireflective layers.

The corrosion problem is aggravated with the use of Al-Cu alloy. The addition of Cu in the aluminum is necessary for reducing the formation of hillocks/voids and for improving the overall electromigration layer of native oxide over it. This must be removed and is done with the use of BCL_3. Once the native oxide is removed, then the Cl plays a major role in etching the aluminum. In order to obtain an anisotropic etch result in near-vertical sidewall profiles, two techniques are used. First, a polymer film is continually formed on the wafer surface during the etch. This film is removed in flat regions by ion bombardment to allow the etching of Al, but remains on sidewalls to prevent undercut. The etch chemistry and conditions control the polymer formation. The polymer on top of the metal needs to be removed so that the etching may proceed. This is accomplished by using nitrogen ions to bombard the horizontal surfaces, thereby clearing the polymer films. Since the clearing of the polymer takes place normal to the incident beam of nitrogen ions, the profile of the metal pattern is strongly dependent on the angle of the topographic step. The metal etch contains an overetch component that is needed to clear any residual metal from the surface. Otherwise, these residual metal filaments may cause shorts and make the devices fail.

Postmetal etch. The postetch process determines the success or failure of interconnect definition. To prevent the wafers from corroding on exposure to the atmosphere, the wafers are passivated prior to being removed from the etcher. This passivation is done by substituting the Cl absorbed in the resist with F from the passivation gas. After this substitution is done, the wafers are removed from the etcher and rinsed in deionized water to remove all the excess Cl. Immediately following this, the resist is stripped using both plasma and organic resist strippers.

The management and control of corrosion has taken the lead role in the Al-based interconnect definition. As new materials are developed to substitute for the Al-based interconnects, new etching techniques will also evolve. As with any other process, the issues pertaining to process control will play a paramount part in the interconnect definition. Corrosion that manifests itself much later in time after the devices have been shipped can be a nightmare for both suppliers and customers.

Metallization applications. There are several applications [Ohba 1991, Kotani 1988, Burba et al. 1986, Gupta et al. 1990, Elliot 1988] where metal films are needed. Table 12-9 shows some of the common metals, their compounds, and their applications. It is being presented here for illustration

Table 12-9 Metallization Application

Application	Material
Silicide/gate	Ti, Co, Ni, W, Ta, Mo
Contact fill	Al alloy, W, Si, $WiSi_2$
Barrier/adhesion	TiN, TiW, Wsi_x, TaN
Shunt	Ti, W, TiW
Primary conductor	Al alloy, W, Cu, Au
Antireflective	TiN, TiW

purposes only. These materials may be used for other applications, and an exhaustive list is beyond the scope of this book. Metallization schemes (the stack composition) may vary depending on the device reliability requirements. In a generic situation, this table may represent a cross-sectional image of the entire metallization stack in a multilevel interconnect system. By choosing a stack approach to interconnect metallization, the reliability of the interconnect is enhanced. Only Al or its alloys by themselves suffer from problems that are detrimental to the reliability of the entire device.

The issues with silicide and contact metallization have already been discussed along with the barrier/adhesion layer issues pertaining to the W contact filling and planarization. Let us examine the metal interconnect stack, specifically the contributions of the barrier layer, primary conductors, shunt layer, and the antireflective layers.

Barrier metal. Barrier metal layers are used to prevent an interactive reaction between two adjacent materials. If no barrier layers are present, the reaction takes place in the form of diffusion of one of the barrier layers. [Nicolet 1978] proposed a classification of the barrier layers. These films may be separated into three groups:

- *Passive barriers.* These are chemically inert against the adjacent layers.
- *Sacrificial barriers.* These materials take part in the reaction and are consumed by it.
- *Stuffing barriers.* These minimize the diffusion along the grain boundaries.

The requirements for barrier metal layers are as follows:

- Good adherence to Si, silicide, oxide, and other metal layers.
- Low contact resistivity is very important since this layer makes contact with the Si or silicide layer.
- The barrier metal films must be able to withstand thermal processing, prevent interdiffusion, and also be easy to etch.

TiN is widely used as a barrier and as an adhesion layer in processes that use W for contact filling. The advantages of combining these are essentially simplicity in processing. TiN is deposited either by reactive ion sputtering or nitridation of sputter-deposited Ti in a nitrogen ambient. The interactions of the barrier layers with aluminum show that TiN has the capacity to absorb Al and react with it, forming an intermetallic compound. This absorption of Al increases with temperature and therefore prevents the diffusion of Al across the barrier.

Primary conductor/shunt layer. Aluminum and its alloys–AlSi, AlSiCu, and AlCu–are extensively used as interconnect materials, primarily because of their high conductivity, ability to form low resistance ohmic contacts with both p^+ and n^+ contacts, and good adhesion to silicon and silicon dioxide. The electromigration performance of the Al interconnect is enhanced by the addition of Cu to Al. As the devices are scaled to submicron regimes, the interconnections may occupy a significant area of the chip and hence the reliability of these is of paramount importance. The requirements, therefore, are changing for submicron interconnects. These need to possess the advantages of Al interconnects, but must be more electromigration resistant, be easy to etch, and overcome the corrosion problems after etch that affect Al interconnects. Several interconnect schemes are in use to make the interconnects more robust. Some metallization schemes [Yamamoto 1987] use W to cover the Al interconnects to prevent hillock formation during thermal stress. Titanium (Ti) layers that are harder and more electromigration resistant than Al may also be used. Therefore, if the Al interconnects suffer from defects that present a current restricting path, the Ti or such layers may still conduct the bulk of the current and prevent the device from failing functionally.

Elegant interconnect schemes not only must make the interconnect system more reliable, but also be easy to etch and overcome the problems associated with the etch. Al and its alloys are etched in a chlorine-containing chemistry that causes corrosion, if the postetch treatments are not managed with care.

Antireflective layer. If the Al surface is highly reflective and coupled to a varying and uneven topography, the reflection of UV light will be uncontrolled and cause CD loss due to unwanted exposure. To prevent this, an antireflective layer is deposited over the metal. This reduces the reflectivity of the surface and helps make the lithography process more controllable.

12.8 Summary

Widespread use of multilevel interconnect technology for enhancing the performance of advanced integrated circuits has hastened process development in several areas. New dielectric and interconnect materials in combination with plasma etching and planarization methods have enabled the successful fabrication of submicron devices. The challenge is to develop and transfer it to high-volume manufacturing process technologies suitable for sub-half-micron geometries.

13. MANUFACTURING

13.1 Introduction

This chapter deals with manufacturing issues associated with integrated devices that use multiple levels of metallization. These issues are quite different for integrated circuits (ICs) that need only a single level of metallization. The fabrication issues are complex in nature and this complexity lies not only in the multilevel technology but also in the manufacturing methodology. Even though a company may have a good product design and process, the real litmus test in seeking a competitive market advantage and share lies in a company's ability to manufacture the products in high volume at the lowest possible cost. This chapter is divided into four segments: manufacturing challenges, process integration, yield analysis, and defect reduction. These topics are examined here in the context of manufacturing advanced ICs, specifically those that use multilevel interconnects.

Manufacturing plays a critical and central role in IC fabrication. The challenges are multifaceted, ranging from advances in design and process to manufacturing practices. The manufacturing challenges may appear simplistic and mundane, but the goal is to take an advanced technology and produce products in large quantities that are safer, faster, cheaper, better and easier to make than those of the competition.

Each process must be developed taking into account manufacturing issues. One of the process integration challenges is in creating a process flow that is easily transferable to manufacturing. The requirements and goals of process integration vary from the process development to production.

Yield analysis is one of the key indicators that conveys the degree of success of the process as a whole and the manufacturing technique in particular. The yield loss of product due to wafer processing is examined, and yield models that help calculate and compare different product yields are also presented in this chapter.

Finally, continuous improvement in both the process and manufacturing procedures drives the defects down, resulting in an increase in profitability. The focus here is on defect reduction methodologies. With such a mix of challenges, the fabrication of ICs is an exciting field of study.

13.2 Manufacturing Challenges

Manufacturing is an exciting field where people, equipment and technology have to be managed for fabrication of products. The challenges this offers are many [Bowers 1990, Cunnigham 1990, Suzaki 1987]. The focus of this chapter is on the manufacturing challenges that multilevel interconnect technology presents. The sequence of steps that are needed to fabricate multilevel interconnects poses the potential for manufacturing bottlenecks and yield depression due to increased defect levels. One of the reasons is because of the multitude of repetitive steps that are required to fabricate interconnects. Effective process transfer methodologies from development to production, coupled with sound manufacturing practices, can overcome most of the problems associated with the process implementation. The *backend* (BE) of a process flow may be defined as the sequence of steps starting after the transistors are formed. That is, the BE of a process flow comprises multilevel interconnects – deposition, planarization, etching, etc. the *frontend* (FE) may be defined as all the steps starting from silicon start to the formation of transistors – active and isolation areas, gate definition, etc.

Overcoming yield problems associated with multilevel interconnects is an extremely important activity because the value of silicon wafers increases as they progress toward the BE. Once wafers leave the fabrication site and go to the assembly site, they increase in value as they are placed in a package and tested. Of particular interest are defects and process deviations caused by interconnect and dielectric steps. These defects and deviations may lead to the loss of a significant number of wafers which have accumulated a potentially high market value. The graph shown in Figure 13.1 shows the cost of identifying defects at various points in the process flow. Obviously, identification and scrapping of defective materials at the start of the process is much less expensive than at the end, which may involve analyzing customer returns for defects and recalling the defective products. It pays (lower cost) to identify and eliminate defects at the source or close to it. As technologies push the level of device integration even higher with use of multiple levels of interconnects, the cost per wafer starts rising. To reduce the cost per die, the migration to large wafer sizes is becoming necessary as more die can be printed on a wafer. However, increased wafer size comes with its own issues.

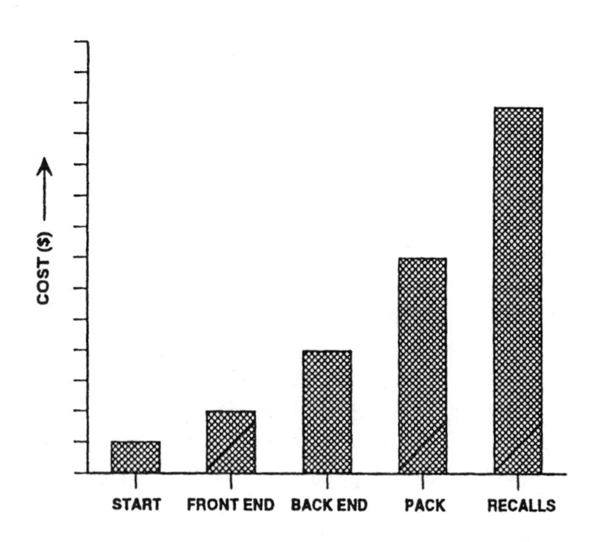

Figure 13-1 Relative cost of identifying defects at some locations in a wafer fabrication flow

13.2.1 Manufacturing methods

There are two methods, proactive and reactive, for achieving manufacturing success. These are not alternatives that one can choose from, but necessities for good manufacturing. Figure 13.2 shows the flow of information from both proactive and reactive activities into the manufacturing process. Each of these methods is a part of the feedback system that is constantly aimed at improving designs and processes so that high-volume manufacturing is feasible with maximum yields.

In the proactive method the manufacturing concerns are incorporated at the design and process development stage. The manufacturing issues are resolved though a constant and regular feedback loop between design, process, and manufacturing organizations in an IC fabrication plant. This interactive learning process enables designs and processes to be developed that can be quickly put into high-volume production. It is only through high-volume production that two important benefits may be realized. The first benefit is an increase in the cycles of learning that may not have been possible during process development due to long throughput times. These cycles of learning shorten the time to understand and fix potential yield limiters. The second benefit of increased volume of production with high line and die yields is the opportunity to penetrate the market ahead of the competition and recover most of the developmental costs in the early phases of product introduction. A successful company will do this before the product transitions from a proprietary to a commodity product.

The reactive mode of operation involves addressing systematic and random excursions that degrade the yield, which is a function of manufacturing operations. Depending on the magnitude and nature of the problem, the manufacturing process may be halted for the same time pending a resolution of the issue. Reaction to excursions can be made easier by using inline and end-of-line monitor data that can flag early indications of the onset of a problem. Experience gained from successfully addressing

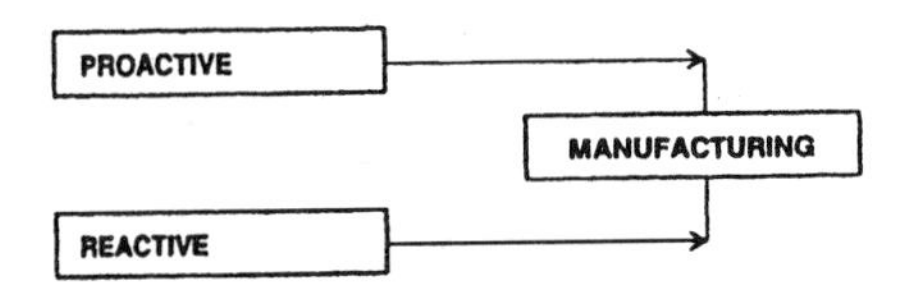

Figure 13-2 Proactive and reactive modes of information flow in a manufacturing system

manufacturing issues may be captured in cause and effect diagrams, troubleshooting, checklists, etc.

13.2.2 Customer concept

Manufacturing of submicron ICs is well placed in a pivotal role; on one hand we have the customer and on the other, we have the company's own profitability and survival at stake. As manufacturing technologies increase in complexity for submicron geometries, the customer needs to be placed in a central position and educated about key operations and products. Such an education helps foster a better relationship with the customer and makes them a partner in the joint progress of both the customer and vendor companies. Figure 13.3 shows a possible scenario of the relationship of various groups in an IC manufacturing company to the customer. This is not to suggest that the customer should deal with a host of organizations. This figure is meant to illustrate that each organization should comprehend the needs of the customer, be it internal or external. Each group should constantly understand the issues of the other groups via an information path that is available for data and information collection and dissemination. Such a relationship is much needed in an application-specific integrated circuits (ASIC) fabrication environment where rapport with the customer is vital to the successful fabrication of the products that meet the customer's needs. In commodity product [such as dynamic random access memories (DRAMs)] fabrication, the relationship may not be as close, but it is still important to ensure that the products and services are performing to standards set by the customer.

Manufacturing directly impacts the profitability of both the customer (on-time delivery of products at competitive cost) and the company fabricating the ICs itself. In Figure 13.3, manufacturing, along with other groups, has a key role to play in enhancing the customer relationship. There are many ways to provide good support to customers; one of them is through education of the fabrication process as a whole. Education brings a better appreciation of the issues facing the vendor.

13.2.3 Technology trends [McClean 1992].

On a macroeconomic scale, the market is continuously emphasizing the decline in IC unit cost. This is especially true in the case of DRAM fabrication where high-volume manufacturing is a significant leverage item in reducing unit cost. Unit cost is further driven down by using larger-diameter wafers and increasing line and die yields. Simultaneously, new generation design and process technologies are being developed to increase

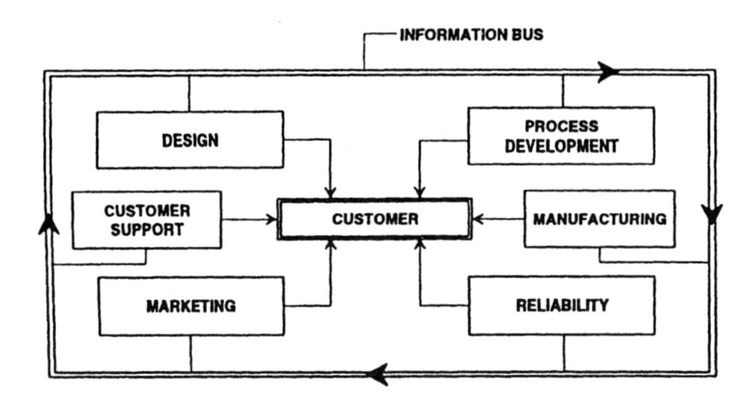

Figure 13-3 A possible configuration of customer interface system with key areas within the vendor organization

the packing or integration density (number of transistors per chip) so that faster and smaller chips may be manufactured in larger quantities.

Packing density. The packing density of transistors has been constantly increasing, primarily due to the need to enhance the application scope of the IC. Figure 13.4 shows an increasing trend of transistor packing density. Defining the interconnects for all these transistors is a challenging task. The interconnect layout must be done using multiple levels of metallization as the die size using a single layer of metallization would be large and more adversely affected by high defect densities. The critical question is, how can we ensure that our parametric and reliability testing are adequate for predicting the functionality and reliability of the millions of transistors in the die? The answer lies more in the methodology of process control and continuous improvement and less on additional testing. The increase in packing density requires new process technologies with ever-decreasing feature sizes in order to extract the full benefit of the product design.

Feature size. The downward scaling of device dimensions enhances the speed and power performance of devices and also allows the increase in the transistor packing density. Figure 13.5 shows the trend of the feature size over time. This diagram illustrates the current production and experimental resolution in terms of device fabrication and prototyping, respectively. Feature size reduction pushes the process technology from one generation to the next. This means that significant process enhancements are required to transfer the designs to silicon (Si). Feature size reduction is limited by fine

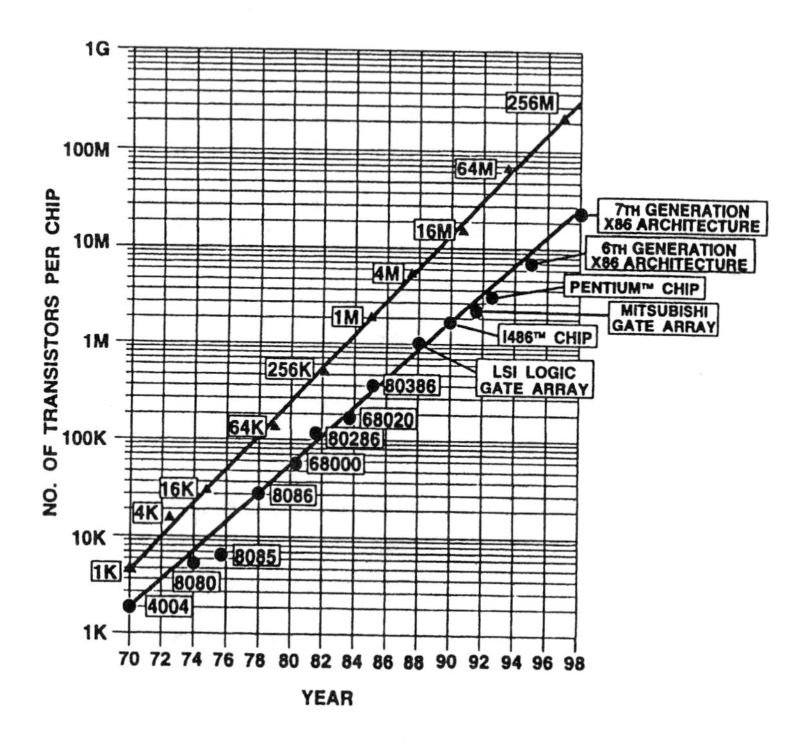

Figure 13-4 Transistor packing density trend [McClean 1992]. (*Courtesy of Integrated Circuit Engineering Corporation.*)

pattern formation of the lithography tools and the ability to develop multilevel interconnect systems. The process technology for each generation is dictated by the speed of development on two fronts – transistors and interconnects. As shown in Figure 13.5, the formation of devices in the sub-quarter-micron regime has been demonstrated. Besides the optimization work for the transistor, one of the key requirements for high-volume manufacturing is the development of interconnect technology. The reduction in feature size presents a whole new challenge toward the definition of multilevel interconnects. Once the process has been developed for both transistors and interconnects, the manufacturing challenge is in producing wafers with high yields, as smaller defects may become potential yield killers.

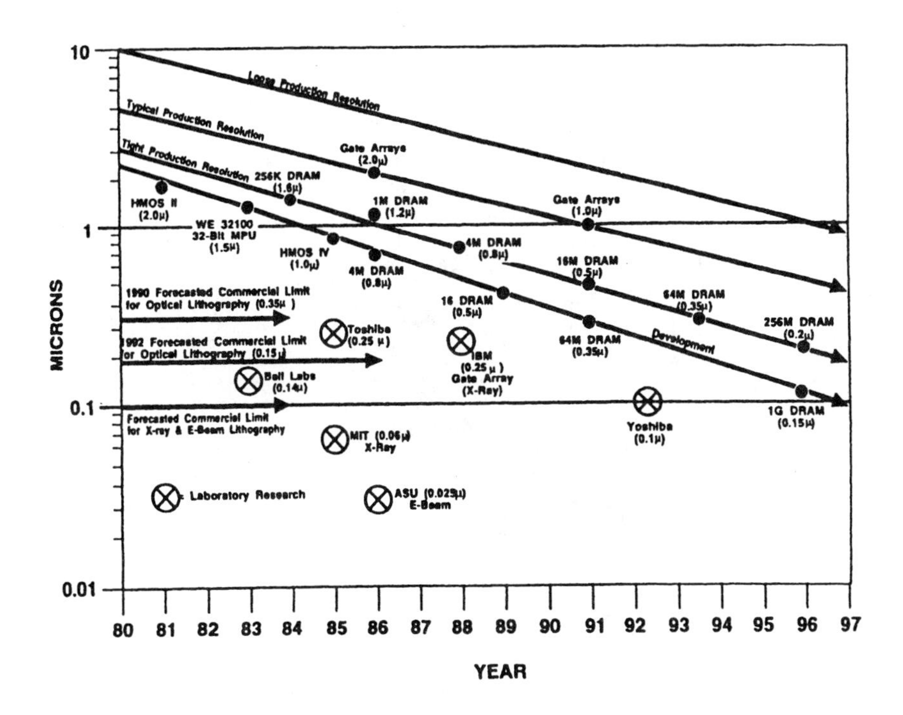

Figure 13-5 Device feature size trend due to scaling [McClean 1992] (*Courtesy of Integrated Circuit Engineering Corporation.*)

Die size. In general, the die size has been increasing with time and keeping pace with the increase in the transistor packing density. This trend is illustrated in Figure 13.6, which shows the die size increase for both the microprocessor and the memory devices. These two device technologies have kept pace with each other. Die size increase is inevitable as transistor packing density increases with each new device generation. For a particular device generation, the die size may be reduced to increase the number of dice per wafer. This is done by design shrinkage and by using the next generation of process technology, which enables the fabrication with smaller feature sizes. The benefits of increasing die size are packing density increase, enhanced circuit layout, and better circuit function.

However, from a manufacturing perspective, die size increase presents several problems as highlighted in Figure 13.7. These problems may be

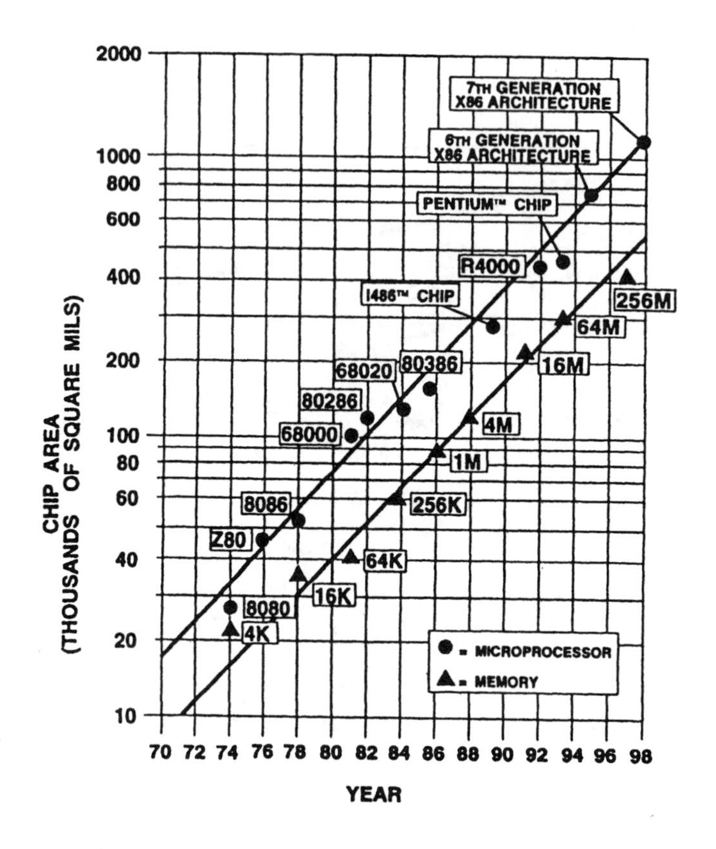

Figure 13-6 Die size trend. [McClean 1992]. (*Courtesy of Integrated Circuit Engineering Corporation.*)

overcome by developing a more advanced technology that may compensate for these side effects. The main problem of die size increase is elevated levels of defects (higher probability of defects falling on die) and hence lower yields.

Wafer size. To reduce manufacturing costs, an increase in wafer size is needed as more dice may be fabricated on a single wafer. The current trend for most advanced IC manufacturers is to transition from 150- to 200-mm wafers. Some of the processing challenges resulting from increasing wafer

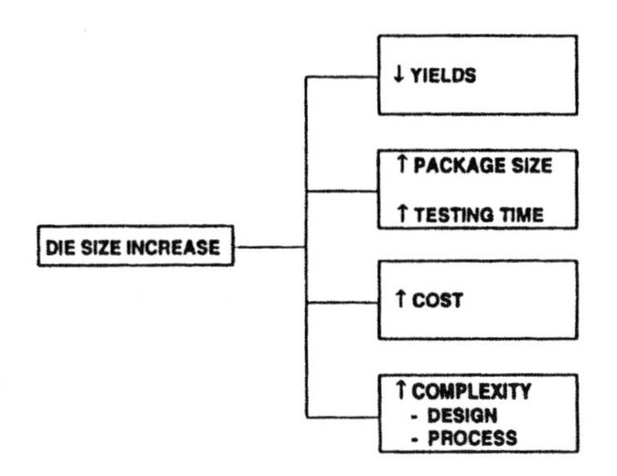

Figure 13-7 Benefits and concerns associated with increase in die size

diameter are modulated by equipment issues in terms of ensuring tight uniformity control across the wafer. Wafer handling and inspection are also more difficult on large diameter wafers.

13.2.4 Process complexity

The manufacturing operation may be divided into two groups, one for transistor formation and the other for interconnect formation. It is interesting to compare these two groups and attempt to understand the complexities of each. The issues listed are the output parameters and concerns that the manufacturing operations must ensure in order to meet process and product specifications. Figure 13.8a and b compares the salient steps in both transistor and interconnect formation processes.

The transistor formation steps are generally batch processing involving large load sizes that may cause major line yield loss should there be a processing problem. On the other hand, some of the interconnect formation groups are in planarization, metallization, and etching. The comparison may be further broken down into materials, temperature, and cleans.

Materials. In the transistor formation process blocks, the process complexity lies in precise control of dopants in the silicon to form the requisite wells and junctions by optimizing the oxidation, diffusion, and implant steps. The thickest film [in a metal-oxide semiconductor (MOS)]

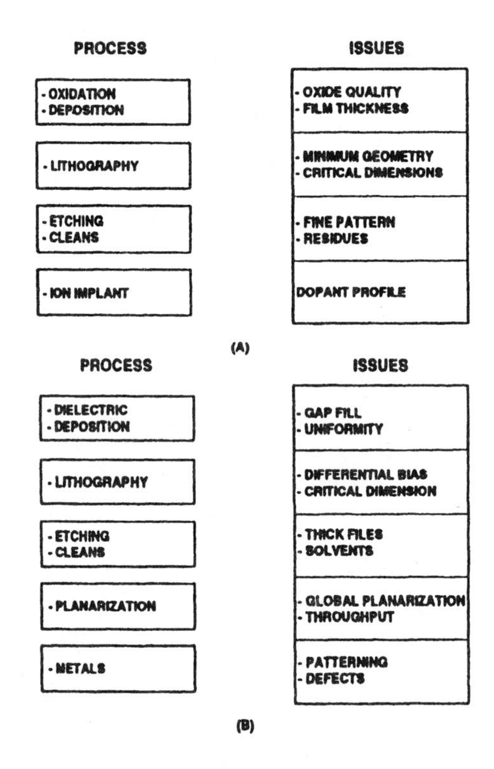

Figure 13-8 Comparison of key steps in (a) transistor formation and (b) interconnect formation

using thin film deposition techniques is the polysilicon layer, and with oxidation it is the field oxide. Even though the feature sizes are small, the aspect ratios are not as large as in the BE or interconnect formation steps so the common problems of filling high-aspect ratio gaps with thin films do not arise. The primary materials that are used in MOS transistor formation are Si, silicon dioxide (SiO_2), and silicon nitride (Si_3N_4). Compare this with the number of materials that are used for the formation of interconnects – several types of dielectrics [oxides, nitrides, spin-on-glass (SOG), and polyimides] and metals [aluminum (Al), tungsten (W), titanium (Ti), etc.]. Each of these materials has different electrical and physical properties that make process development and control a challenging task. The management of the materials is therefore critical for the manufacture of ICs with multiple levels of interconnects.

Temperature. High temperature is a major constraint in all of the interconnect formation steps as Al and its alloys are widely used as interconnects. Pure Al melts at 660°C and this temperature starts decreasing as Al alloys with Si and/or copper (Cu) are formed. The Al/Si eutectic temperature is around 577°C and the solubility of Si in Al rises as the temperature increases [Wolf and Tauber 1986]. Therefore, Al-Si alloys are used to prevent junction spiking and the post aluminum temperatures are restricted to around 400°C. An increase in temperature also causes the Al to relieve stress in the form of hillocks. Increase in temperature induces a stress in the metal due to a difference in the expansion coefficient of the metal and the silicon. Hillock density may also increase with an increase in impurity content (nitrogen). Temperatures in the dielectric deposition are also controlled with the usual operating temperatures around 300°C to prevent the formation of hillocks. Unlike the interconnect steps, transistor formation steps (oxidation, diffusion, and deposition) are done at elevated temperatures. Temperature effects are well characterized and understood for thin films and dopant distribution in silicon.

Cleans. In addition to the number and nature of materials, the wet chemical treatments are different for each group. For the transistor formation group, acidic cleans do a reasonable job removing resist residues and other contaminants. However, for the interconnect cleans, nonacidic cleans are necessary in order to prevent metal attack. Organic resist strippers are widely used in addition to plasma ashes to strip resist and clean the surface. These cleans are not as efficient as acidic ones that are used for the transistor formation steps.

These differences present a different set of process integration challenges for the two groups. However, from a manufacturing perspective, the issues between the two groups are similar except when it comes to metallization, which is a critical operation in the manufacturing flow. The manufacturing issues stem from the repetitive nature of the interconnect processes. The financial value of the wafer increases with each interconnect level. Misprocessing at any of these stages causes a significant loss of time and money.

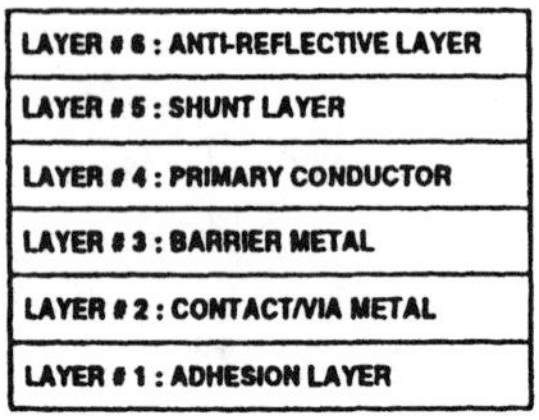

LAYER # 6 : ANTI-REFLECTIVE LAYER
LAYER # 5 : SHUNT LAYER
LAYER # 4 : PRIMARY CONDUCTOR
LAYER # 3 : BARRIER METAL
LAYER # 2 : CONTACT/VIA METAL
LAYER # 1 : ADHESION LAYER

Figure 13-9 Identification of components in a typical metal interconnect

13.2.5 The metal loop

As the interconnect levels increase, the metallization complexity in terms of manufacturing logistics increases too. Figure 13.9 breaks down the actual metallization steps that are required to form one level of interconnects in a typical multilevel interconnect system. A process that has three levels or four levels of interconnects consists of several metallization steps. In Figure 13.9 there are six steps that form one interconnect level. So, for a three-level multilevel interconnect system, there are 18 metallization steps. Inclusion of source/drain and polysilicon metallization will increase the total number of metallization steps still further. This assumes that the metallization scheme is Al-based, as barrier and shunt layers are needed to make interconnects robust. Multiple metallization steps could constrain the flow of material. The bottlenecks may be equipment or process related. Some relief may be obtained by using common metallization steps for various applications. For example, the adhesion layer in a W plug process may be TiW and these films may also be used as barrier and antireflective layers. This enables the same equipment set and process to be used over and over again. Such a methodology in process development makes manufacturing more efficient.

Line balance is essential to ensure a smooth flow of material through the manufacturing line. In a short-cycle time environment, line balance requires that the various operations be synchronized, as they normally run at different speeds. The interconnect steps are a mixed bag of single-wafer and batch process steps which are fed by the FE operations which are mostly batch. The situation is further compounded by a multitude of repetitive non-value-added steps such as inspections for defects, and machine and process monitors in the metallization steps. Therefore, along with the reduction in cycle time, a careful evaluation of the non-value-added steps is required. An effective utilization of the monitors and inspections is needed to not only

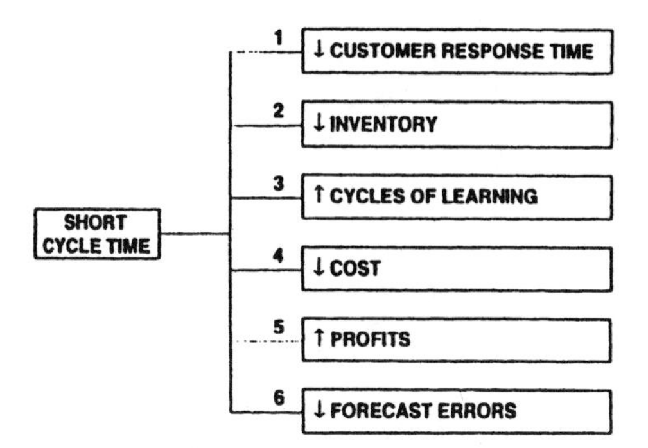

Figure 13-10 Benefits of short manufacturing cycle time

obtain all the necessary information for checking the status of the process but also to improve the flow of materials through the line.

13.2.6 Cycles of learning

In both development and high-volume manufacturing the benefit of shorter cycle time cannot be overstressed. It is the most important manufacturing indicator which establishes the success of the manufacturing operation. Short cycle times increase the number of cycles to learning or information turns and therefore provide an opportunity to reach precise excursions quickly. What is cycle time?

Cycle time or throughput time is the time required for material to move from the start to the end of the flow. It is a measure of how quickly the work-in-process (WIP) moves though the line. Each piece of equipment has its own throughput time and the sum of these times plus the material handling times plus inspections make up the theoretical throughput time of the manufacturing line. In practice, however, cycle time is influenced by line yield at each step, the WIP levels, and manufacturing practices. Cycle time may be expressed by a simple formula, as shown in Equation 13.1.

$$\text{Cycle time} = (\text{WIP inventory}) / (\text{processing speed}) \quad (13.1)$$

The advantages of having short cycle times are many and some of the salient ones are given in Figure 13.10 (a complete listing of the advantages is beyond the scope of this book).

Customer. This consists of benefits 1 and 6. A decrease in the cycle time will help the customer by predicting the product delivery schedule more accurately and by providing a quick response. With shortened cycle time, problems become visible and the reaction to fix them can be started in a timely manner.

Profits. Benefits 2, 4, and 5 may be considered to be part of this group. One immediate advantage of achieving short cycle time is the reduction in inventory and its carrying costs. Just-in-time manufacturing concepts may then be applied to further leverage the manufacturing operation in securing the gains of reduced inventory levels. The relationship between cost and cycle time needs to be explored a little further. As we continue to decrease cycle time we are able to reduce costs to a point. There is a point when further reduction in cycle time begins to increase cost and is no longer attractive.

Learning. One of the key benefits of short cycle time is the opportunity to increase the cycles of learning. This benefit is vigorously sought by engineers to understand the cause and effect of various processes and to determine the effectiveness of process improvements. The concept of a cycle of learning (CL) is very important to yield improvement. A cycle of learning may be defined as the elapsed time from a process change until the results of the change are analyzed. Expressed mathematically, cycles of learning (CL) may be defined as:

$$CL=WD/CT \tag{13.2}$$

where WD= workdays in a year
CT =cycle time in days

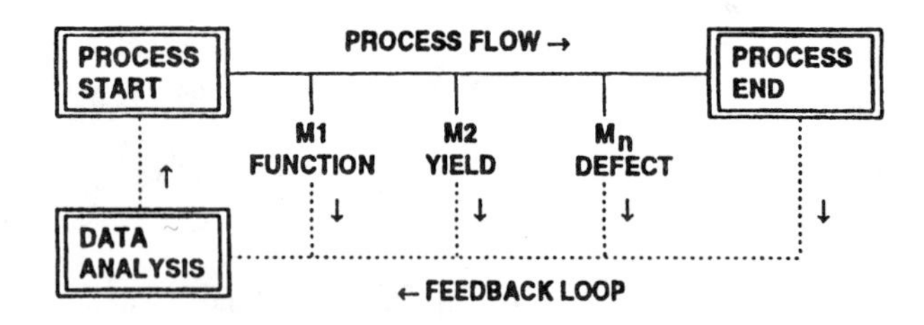

Figure 13-11 A process's health sampling feedback system

Increasing the cycles of learning in a manufacturing operation provides the opportunity to identify problems and rectify them faster. With the increasing sophistication of processes, the number of steps is increasing and this tends to make cycle times longer.

Cycles of learning may be increased in two ways, first by reducing the overall cycle time and second by sampling the process at regular intervals and reacting to the data before a problem reaches the end of line. Let us take a closer look at CLs as a result of increased sampling by looking at a multilevel metal process technology flow with n layers of metal. Figure 13.11 shows a feedback loop in a n-level metal process flow. In this figure, the process begins with the Si wafers being started for the various products. After completing their inline processing, the Si wafers move to test where die yield is determined. As this process uses multilevel interconnects, it may be segmented into several sub-processes and the learning activity may be tailored for each sub-process. In a multilevel metal system, let the final metal layer be M_n. The process flow may be sampled at M_1, M_2 M_n interconnect levels before the lot reaches test for yield evaluation.

Transistor functionality may be tested once the first-level interconnects (M_1) are made. Testing samples at this location provides a quick and early indication of the transistor functionality. This is a good location to check for contact resistances, threshold voltages, etc. To speed up the learning process, special short process flows may be created that terminate at M_1. The short loop process flow can take advantage of an inherent low cycle time of the manufacturing line, and further increase the cycles of learning. Experiments may be defined on these short loops in order to characterize process specification windows associated with some of the critical processing steps.

The above concept may be extended to other levels of interconnects. To check circuit functionality in some devices, the wafers may be processed on a different flow and terminated after the second interconnect level (M_2) has been defined. This provides an early indication of any yield problems. This is an excellent way to check the yield impact of a process change; for example, on static random access memory (SRAM) yield vehicles that may be fabricated with two layers or more of metallization. Low-yield analysis may be performed on such material and valuable information on the failure modes may be obtained well before a fully processed lot with all levels of interconnects makes it to the end of the line. In addition to this, separate process flows may be created to check the continuity, via resistances, etc. As each interconnect level is similar to the previous one in a multilevel interconnect system, the defect generation sources may be studied if the wafers are inspected at the last interconnect level and then sent on for end-of -the-line testing. Inspection for defects after the final layer does provide a cumulative defect count. This provides an opportunity to correlate the visual defect density to the electrical defect density (yield), which enables fast feedback for reaction. All wafers are tested twice: first, to check the

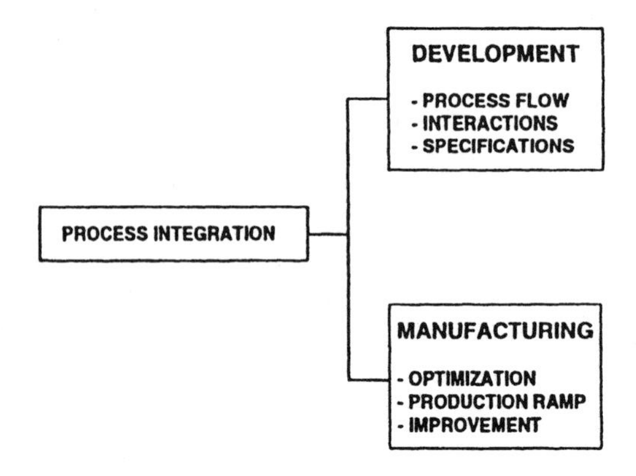

Figure 13-12 Process integration challenges during process development and high-volume manufacturing

electrical parameters and second, for the circuit functionality. Information from these two tests helps determine an appropriate failure analysis, if needed. The bottom line is to increase the cycles of learning through aggressive management of cycle time reduction, creative process sampling, and monitoring.

13.3 Process integration

Process integration is one of the key functions in the process development and manufacturing of ICs. There are many integration aspects when fabricating ICs with multilevel interconnects. Process integration focuses on the health of the process as it migrates from development to high-volume manufacturing. This process is highlighted in Figure 13.12 where areas of focus are listed in the blocks.

13.3.1 Development

During the process development phase, the process integration is concentrated on defining process flows, characterizing new processes, studying process interactions and addressing systematic process problems. An example of process interaction is the stress in thin films which must be carefully managed. Mismatch of stresses between films may lead to film delamination or other catastrophic failure. Once interactions are understood, the process must be controlled within specification limits that may be based on product requirements and on the process variability.

13.3.2 Manufacturing

As shown in Figure 13.12, key integration issues in high-volume manufacturing are different from process development. Successful transfer of a process from development to high-volume manufacturing depends on its maturity and the discipline in technology transfer. Process optimization and improvement may be done after process transfer and are required for the reasons given below.

Machine matching. Several pieces of equipment may be used for any process step. It is absolutely essential to ensure that each of these machines are matched so as to obtain a tight distribution of the output parameters. The objective is to drive output electrical parameters closer to target for all machines and reduce the variation between them.

Process envelope. The process envelope or window is the region in which the process may operate without jeopardizing products. During process development, process window experiments can be used to characterize a process. However, during high-volume fabrication the process envelope may change due to volume effects. As an example of volume effect, consider a process flow that may have been developed with a short time delay between borophosphosilicate glass (BPSG) deposition and densification and reflow, avoiding crystal formation. In volume production, the characterized time between BPSG and densification may not be adequate. Cycle time and manufacturing practices may require a wider time limit between processes. Therefore, reevaluation of the process envelope to provide a larger time window for operation may be needed if the process must be improved and made suitable for high-volume manufacturing. In some cases, process improvement is possible and in others it is difficult. It is important not to confuse process optimization (centering the process and verifying the process envelope) with re-engineering/development of a process. If required, re-engineering/development of a process may enhance the yield or reliability of the product, and should be done carefully. In high-volume manufacturing, the risk to the product is far greater than tin development, mainly due to the larger amount of material that is being processed. Process problems due to inadequate characterization in development will cause a significant loss in material, time, and revenue in production.

Problem investigation. The process integration function in a manufacturing environment is more investigative in nature as opposed to discovery/invention in process development. Investigative engineering requires a thorough understanding of the process and process interactions. The key is in problem solving. Failure to quickly and successfully address production issues may result in downtime and substandard product. The lessons learned from various problem-solving opportunities enable designers and next-generation technology developers to improve new product designs and processes. Through investigative engineering both proactive and reactive components of manufacturing may be exercised. The reactive part translates into a process fix while the proactive part may translate into a modification of design or process. It is for this sole purpose that the dialogue between designer, product, process, and reliability engineers is extremely important. What tools are needed for investigative engineering? There are many tests, and customer feedback. Process monitors are widely used to monitor the equipment and the process. For example, these monitors may include thickness or critical dimension checks. Process monitors may or may not flag a device problem, but electrical tests do provide some level of confidence that the product has met specifications as it leaves the factory.

There may be a tendency to increase the number of monitors that check various conditions of the process and equipment. But this poses a problem, as monitor wafers are processed at the expense of product wafers. Moreover, the results of the monitors may gate movement of product wafers and hence increase cycle time, which results in reduced CLs. It is important to evaluate the monitors across the process and optimize their function as the factory ramps to process a larger volume of product.

13.4 Yield Analysis

The aim is to make high-yielding wafers and reliable dice. In an ideal situation the yield of a properly fabricated wafer would be 100 percent. In practice, however, this is not the case and the yield varies depending on the maturity of the process from a few percentage points to near 100 percent. The mechanisms of yield loss are many and the focus of this section is to examine how multilevel interconnect processing effects yield [Sze 1986, Moore and Walker 1987, Moore et al. 1987, Hoyt 1987, Parks 1990, Parks et al. 1989, Winter and Cook 1986, Stapper et al. 1983, Dimitrijev 1988, Lea and Bolouri 1988, Flack 1986, Feldman and Director 1991, Maly and Strojwas 1991, Michalka et al. 1990, Dance and Jarvis 1992].

13.4.1 Yield components

The causes of yield loss fall into two main defect categories: systematic and random. During process infancy, systematic defects are the main contributors to yield degradation. As the process matures, the random component becomes the focus of defect reduction as systematic defects decrease. Figure 13.13 shows defect reduction over time for any process. This may be attributed to progress in addressing the systematic and random defects that degrade yield. Also shown in this figure are two other curves, one representing the systematic component and the other the random. To a large extent, systematic yield degradation is addressed during the process development stage. The process should be transferred with a minimum of systematic defects. However, as process transfer from development to manufacturing may expedited by market pressures, the burden of reducing both types of defect modes lies with the manufacturing organization, when production is being ramped to meet demand. Once the majority of systematic defects are addressed, yield and reliability are limited by the inherent random defect component of the process. The random component is always present even in mature and high-yielding processes. At some time after the process has demonstrated maturity (high line and die yields with no reliability issues), a decision has to be made whether to invest resources to

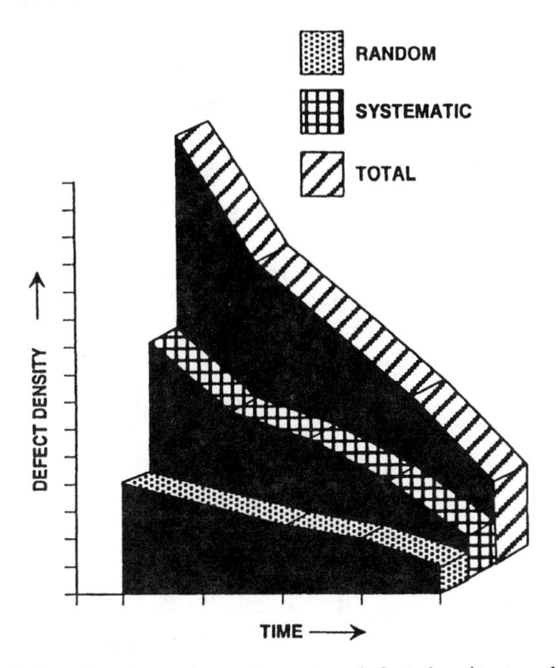

Figure 13-13 Total, systematic, and random defect density trend

bring defect levels down further or to switch to the next-generation process which may offer a yield advantage. With each new generation of process technology, defect specifications for selecting a piece of equipment are tightening.

Systematic yield loss. Figure 13.14 shows a wafer with a grid of dice that have been tested. On this grid "r" represents dice that have failed due to the presence of random defect and the square dot pattern represents dice that have also failed, but due to a systematic problem. Large failing areas suggest systematic problems that may be due to some process step or an interaction of steps. An example of systematic yield loss is the mismatch of uniformities of adjacent process steps. Several systematic yield problems may be attributed to noncomplementary thin-film deposition uniformity profiles and the subsequent etch uniformity profiles, even though for the individual process steps uniformity may be in the specification. Consider a thin-film deposition process that exhibits thickness uniformity discrepancy between the center and the edge of the wafer, with the center being thinner. Consider also the subsequent plasma etch profile that has a high etch rate in the center as compared to the wafer edge. When the two processes are combined in a process flow with the etch process following thin-film deposition, the remaining film thickness will be thin in the center as compared to the wafer edge. When the two processes are combined in a process flow with the etch process following thin-film deposition, the remaining film thickness will be thin in the center as compared to the wafer

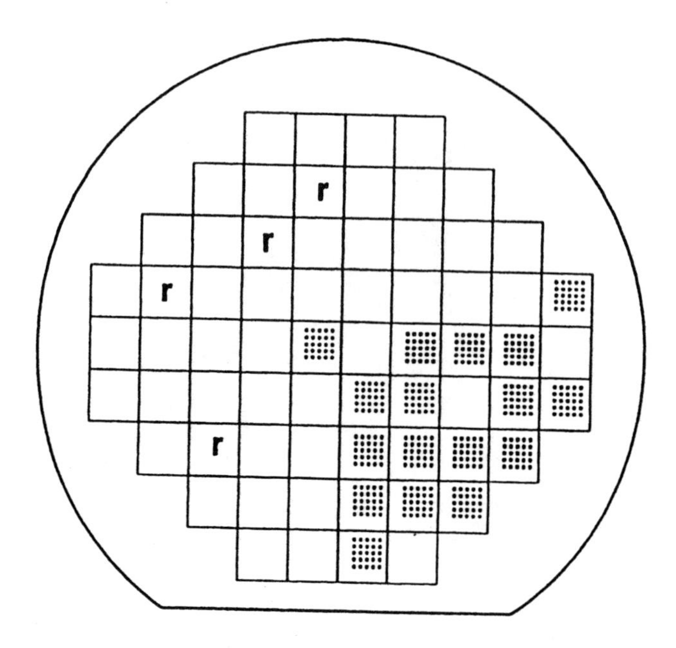

Figure 13-14 An example of a wafer map after completion of yield testing showing die failure due to random and systematic problems

edge due to overreaching. This could result in severely thin areas in the center of the wafer that may lead to processing problems, as shown in Figure 13.15, where affected die are denoted by the square dot pattern. Several combinations of uniformity profiles exist and their interactions may or may not produce an adverse yield condition. Such a systematic problem may be fixed by adjusting the uniformity profiles of upstream and downstream process steps so that they do not interact in a negative way to create an overetched or underetched situation, as described in the above example.

Random yield loss. Yield loss is due to a random occurrence of defects, then failure will be randomly distributed across the wafer as shown in Figure 13.14. Random yield losses are more difficult to address and isolate to a particular processing step. Several models exist to evaluate yield loss due to the random yield component. The impact of multiple types of random defects on the yield is also part of some of the models.

13.4.2 Yield loss.

The cause and effect diagram for yield loss is shown in Figure 13.16. This diagram also highlights the impact of multilevel interconnect

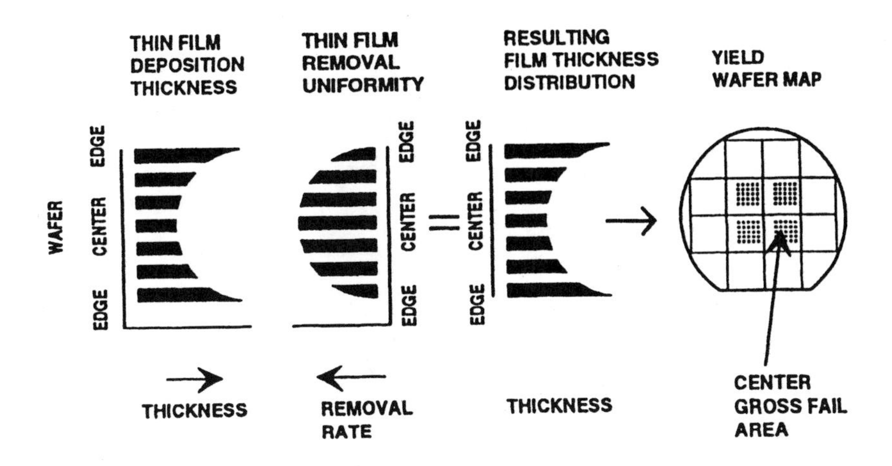

Figure 13-15 Impact of destructive interference of deposition and etch process uniformities on yield

into four groups. As will be more evident in the defect sections, processing on yield. To gain a better understanding of the BE issues that could cause yield degradation, the cause and effect diagram has been divided classification of defects into systematic or random is not always clear-cut. But, irrespective of this difficulty, the cause and effect diagram highlights the various defects that degrade yield in the BE of the process. Particles are a real problem and affect all aspects of processing from the beginning to the end of the process flow. It is to be hoped that when a product is ready for high-volume manufacturing, most of the product issues will have been resolved. In some cases, problems are discovered after a product has been put into production. Poor manufacturing practices may cause yield losses.

Yield degradation may be classified as either catastrophic or parametric in nature. Catastrophic failures are those that make the circuit fail completely and the yield is zero or close to it. Parametric failures are those that do not drop the yield significantly, but change the device parametrics.

Table 13-1 Catastrophic and Parametric Failures

Defects	Catastrophic failure	Parametric failure
Systematic defects	Opens Shorts	Step coverage High resistances
Random defects	Extra material Missing material	Particles Pinholes

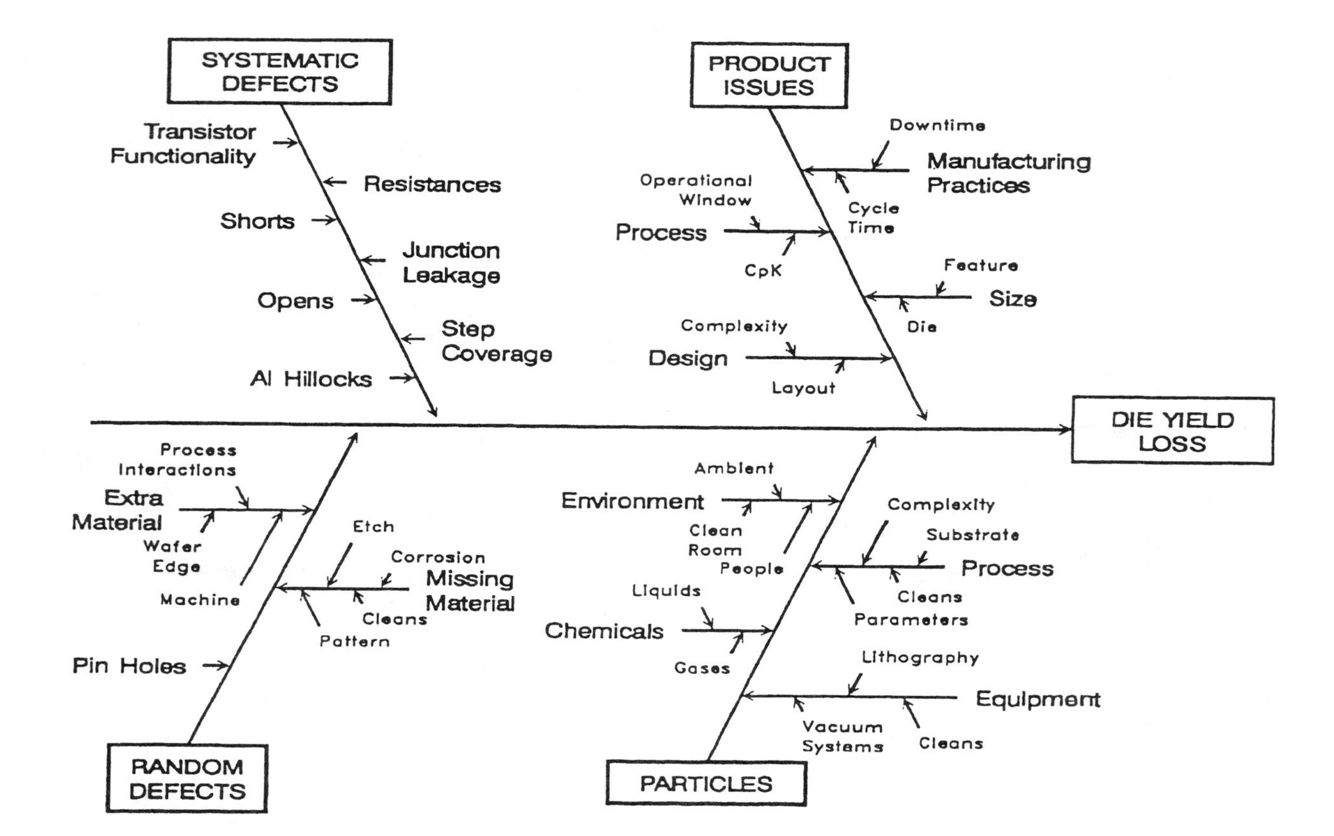

Figure 13-16 Cause-and-effect diagram for yield loss

Such a change, for example, could be a loss in speed of the device due to a process marginality or high resistances. Table 13-1 looks at the issue of yield loss from these two types of failure modes. The examples are for illustration only and this is not a complete listing of each category. Classification of yield killers is not always easy; at times the cause may be systematic and at other times random in nature. Particles are a type of defect that may fall in either the systematic or random categories depending on the source of the problem and downstream processing.

13.4.3 Yield enhancement

Upfront investment of time and effort to understand and quantify the yield impact due to design, process, and manufacturing techniques helps in the long run in achieving profitability. Short-term fire-fighting and Band-Aid solutions may solve the immediate yield problems, but these problems may remain a nagging source of low yield over the long haul. Let us look at the single most important aspect of each of the three components to wafer fabrication– design, process, and manufacturing. Table 13-2 highlights some of the most important aspects for each of the three components.

Defect-tolerant circuitry. As we move to higher and higher levels of device integration, there is a demand for designs with a high degree of defect-tolerance architecture. Memory chips have moved in this direction by incorporating redundant circuits; this has been done at the expense of some Si real estate. Should a particular bit, row, or column fail due to a defect, additional columns or rows may be activated to compensate for the one that is lost. For example, this may be done by using fusible links in the circuit. Reconfiguration may be also done through the use of lasers for altering the wiring layout.

Non-memory architectures are constrained to reroute internal communication paths around defects to make them more defect-tolerant. Several defect-tolerant designs have been proposed that go into greater detail on this subject. To design circuits that are defect-tolerant, a good

Table 13-2 Key thrusts for achieving high yields

Area of yield improvement	Key modulator	Direction of Change
Design	Defect-tolerant circuitry	Enhance defect tolerance
Process	Variation	Reduce process variation
Manufacturing	Cycle time	Decrease cycle time Increase cycles of learning

understanding of the defect and yield models is necessary. The relationship between the yield/defect modeling and defect-tolerant designing is highly complementary.

Process variation. In the case of processing, the single most important thrust is reducing the variation for every process, once it has been developed and is ready to be used for production. To be used for production, a process must be deemed to be repeatable, predictable and controllable. Continuous improvement is a key to maintaining and improving yields. The reduction of process variation needs to happen across the board for all the process steps. Let us take an example from the contact formation step and try to understand the effect of the variability of the processes on the final result – contact resistance. The contact formation step is influenced by the field oxide, polysilicon, and BPSG thickness in a typical complementary metal-oxide semiconductor (CMOS) process. The variation in each of these processes has a cumulative effect on the overall thickness of the step. In worst-case situations, the contact may be formed with a depth equivalent to this step. The contact etch process itself is influenced by its own variability, namely, the etch rate. It is possible for wafers to experience a worst-case scenario of film thickness and etch rate. In order to compensate for such an event, the etch process contains an overetch component of time to ensure that all contacts are opened. This cumulative effect is part of the reason for the variability in contact resistance. If one considers the multitude of interconnect steps, the effect of process variation is significant. This makes yield forecasting all the more difficult and unpredictable.

13.4.4 Yield models

How can we predict of a particular device? What should the yield target be for each quarter? These are just a sample set of questions that must be asked in order to improve the yield for a process and to exceed the targets. Note that the yield models cannot forecast the yield but help determine the current status and set realistic goals for different devices. Most yield models are based on random defects impacting yields.

Total yield. Systematic and random defects together determine the overall yield of the process. The systematic yield loss is not dependent on critical area, unlike the random yield loss component. Critical area may be considered as that area of circuitry that is susceptible to failure due to defects. Therefore, a first-order yield model consists of two components, one dependent of the die area and the other not. This may be expressed as

$$Y = Y_s Y_r (A_d, D) \tag{13.3}$$

where Y is the total yield, Y_s is yield due to systematic defects, Y_r is the yield due to random defects, A_d is the area of the die, and D is the defect density.

In any process flow, critical masking steps may easily be identified if they affect density contribution. Such a classification helps in determining the random defect density for a select number of layers instead of all the layers in a process flow. Once the critical masking steps have been identified, the number of critical operations may also be identified that are part of the critical layer.

If n is the number of critical operations in a process flow, then the yield may be expressed as:

$$Y = Y_s \sum_{i=1}^{n} Y_{ri}(A_{di}, D_i) \tag{13.4}$$

For accurate yield modeling, both the systematic and random yield components need to be modeled. There are several random yield models and the yield statistics associated with each are described extensively in the literature [Moore and Walker 1987, Moore et al. 1987, Winter and Cook 1986, Stapper et al. 1982]. All the models agree with one another when the yields are high and the die area is small. But with ever increasing die sizes, the yield models and associated statistics need to be modified and a model must be chosen that best fits the empirical data.

Random yield models. A key component to the yield models is the magnitude and distribution of the defects. The models described below address one or more of the following defect distributions:

- Uniform distribution
- Nonuniform distribution
- Radial distribution

In all the models discussed here, Y_r is the yield due to random defects, D is the defect density, and A is the critical area.

Uniform distribution. The Poisson model, given by Equation 13.5, is well suited when the defect density is uniformly distributed across the wafer. In real life this is not the case, but within a small critical area the defect density may be considered as uniform. Under this condition, the yield obeys the Poisson equation:

$$Y_r = e^{-DA} \tag{13.5}$$

That is, the yield will exponentially decrease if the defect density, the area, or both increase.

Nonuniform defect distribution. An important enhancement to the above model was based on the assumption that defect distribution is not uniform, even though defect density, D, is constant. Let f(D) be a function that describes defect distribution. As the model takes into consideration varying defect distributions, the yield may be expressed as

$$Y = \int_0^\infty e^{-DA} f(D)dD \tag{13.6}$$

where f (D) is the probability density function

$$\int_0^\infty f(D)dD = 1 \tag{13.7}$$

The above yield equation was developed further and several distributions were evaluated under various conditions. It has been shown that the triangular probability density function best describes the yield that is a function of a nonuniform defect distribution. Therefore, the yield model is given by

$$Y_r = [(1-e_{-DA})/DA]^2 \tag{13.8}$$

The Seeds model given by Equation 13.2 also assumes that the defect density is nonuniform across the wafer and from wafer to wafer. Here the probability density function is exponential and is best expressed by the following equation:

$$Y = e^{-\sqrt{DA}} \tag{13.9}$$

Radial defect distribution. Several defects show a radial pattern in their distributions due to some processing steps. For example, particle distribution in a spin, rinse, and dry system may exhibit a radial pattern as the wafers are rotated around the axis of the boat at high speed. In other words, the defect density is different at the center as compared to the edge of the wafer due to defect clustering. In view of this, some yield models take into consideration this radial variation in defect density.

Example. Let us use one of these yield models to determine the defect density (due to particles) at some critical backend step. How do we determine the particle density requirement for the process that we plan to use is, for each critical module in the flow? The answers to the question will help determine the target goals for particle reduction once the systematic yield killers are identified and eliminated.

A simple step-by-step methodology for determining the particle density requirements for a process is shown in Figure 13.17. This is a four-step solution to the problem and may be extended to other similar problems. One of the requirements for any yield model is to be able to break up the yield for each critical masking step. Such a breakup in yield by masking steps enables a focused yield enhancement as we can then determine the defect budget for each critical layer.

Step 1: Identify critical process steps that can cause yield degradation due to particles. In the transistor formation portion of a typical CMOS device, there are three critical steps:

- Isolation area definition
- Active area definition
- Gate definition

For a two-level metal system the following are the critical steps:

- Contact definition
- First-level interconnect definition
- Via definition
- Second-level interconnect definition

The number of critical steps will increase as we have more levels interconnects.

Step 2. Determine the critical surface area for each of the critical steps. Calculation of the surface area is important for understanding how much area may be affected by particle contamination. For the interconnect layers, the area is the product of critical spacing and the length of the interconnect with this spacing. For the contact and via layers, the area is total contact and via area, respectively.

Step 3. Determine the defect density. One of the yield models may be used for determining the defect density based on the empirical experience with the yield model, process maturity, and the yield improvement progression. Let the number of the process step components as defined in step 3 be k. Then the overall yield, Y, may be defined as:

$$Y = \sum Y_k \tag{13.10}$$

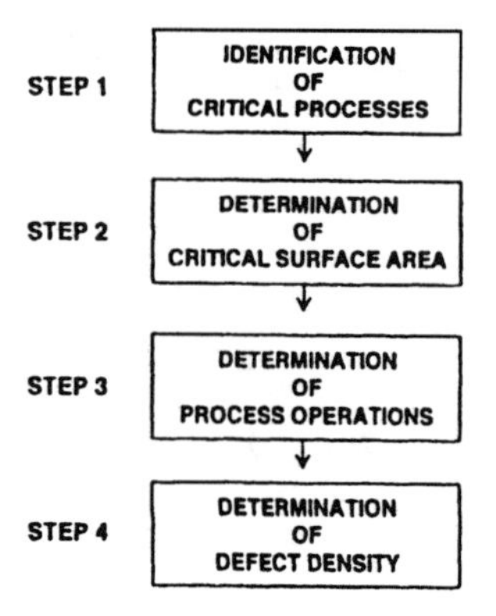

Figure 13-17 Steps for determining particle density requirements

Using one of the yield models, the defect density may be calculated for various levels of yields. As the process matures and higher and higher yields are realized, the appropriate defect density requirements for each critical process may be used as goals for the defect reduction activity.

Using the defect density number calculated from the above method, it is simple to determine the yield loss by equipment. The focus of defect reduction may then be to realize a quicker gain in yield. Note that the above step-by-step method can become elaborate depending on the assumptions used for defect density determination.

13.4.5 Yield vehicle

A yield vehicle is often used to characterize and modify the process as necessary to achieve higher yields. In order to evaluate all the key electrical output parameters, a test product is used. This must contain test structures that can monitor parametric and catastrophic yield loss. For parametric yield loss, we can use the test structures and SRAM chips, both may be printed on the wafer in a number of ways; one way is an alternating array of test and SRAM dice.

SRAM. The use of SRAM chips as yield vehicles in combination with test chips has proven to be a viable yield monitoring vehicle. Such a vehicle is necessary, especially when the process is being developed for logic or custom IC applications. The SRAM has found a wide application as a yield vehicle because of several advantages that are illustrated in Figure 13.18. Owing to the repetitive pattern of SRAM circuitry, failure analysis is simplified. The failure analysis is accomplished in two ways. First, bit mapping can be used to isolated the failing rows and columns. Identification of the failing bits, rows, or columns helps identify the failing location for a thorough failure analysis by removing layer by layer, if necessary. If the bit mapping capability were unavailable, then failure analysis would be time-consuming and would rely on a blind approach. Other memory devices may also be used as yield vehicles.

As the intent of the SRAM chips is to flush most of the systematic problems, process development without such a chip is a definite handicap. Inline visual inspection for defects is much easier as the circuit pattern is similar across the die and defects are easy to locate. The defect location may be correlated to end-of-line failure analysis if the defect or its effects are still present. The SRAM chip, like any other yield vehicle, is only good for visual defects. Nonvisual defects, or rather their impact, may be analyzed by the various electrical test structures that should be used along with the SRAM test chip. Both of these together provide a powerful tool for debugging yield problems that may be either parametric or catastrophic in nature. Once the process is fairly well characterized, logic/custom devices may then be run with reasonable confidence that the yield loss on these devices is mostly due to random defects or design flaws. The SRAM yield vehicle establishes the baseline yield and defect density levels that other products may be compared against and improved on. Let us examine the test structures that complement the SRAM chip for yield analysis.

Test structures. Once the process flow has been defined, the next step is to develop a method of testing the various device parameters to make sure that the flow is basically sound. The primary function for the electrical testing of various structures is to identify processing problems and initiate fixes as appropriate. In addition to this, electrical test structures are used for yield enhancement and as reliability screens. Test structures are designed and placed in specially designed test chips. The test structures are used for

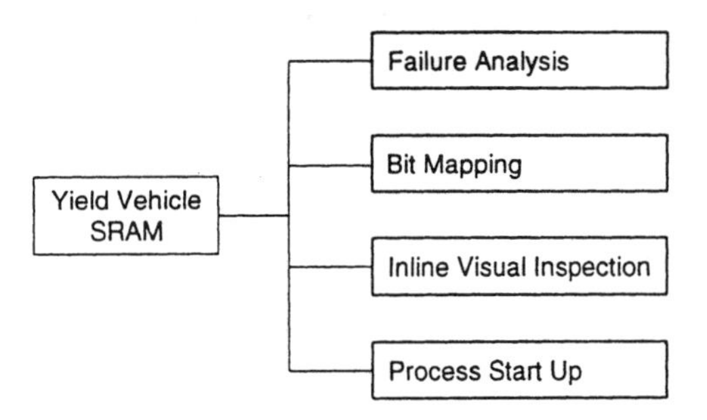

Figure 13-18 Advantages of using SRAM as yield vehicles

understanding transistor functionality, verifying design rules in Si, and monitoring the process (like contact resistance) both from a yield and reliability perspective. Due to Si real estate limitations, it may not always be possible to incorporate all different kinds of structures on the die. Here again, a compromise between information needed for process development and process sustaining needs to be made. Products that serve as yield vehicles may contain an elaborate matrix of structures that can provide information on various processing issues.

The electrical test structures may be divided into four distinct groups, as shown in Figure 13.19. This figure shows the various types of electrical structures that may be placed on a wafer. These four types or variations of these are aimed at detecting a cross section of process problems and defects. For multilevel interconnects there are not as many electrical test structures as required for the transistor formation. We can use structures from each of the above four categories to determine the various issues associated with multilevel metallization. The most common information that is pertinent to interconnects and that is extracted from the electrical test structures is shown in Table 13-3.

Let us review the three resistor test structures (contact resistance, opens or shorts, and misalignment) and examine how the information may be used to monitor the process to gain an idea of the scope and nature of electrical testing of these structures.

Contact resistance. Contact resistance monitors are used extensively to debug and monitor the process. These resistance measurements are generally made by using Kelvin structures with four pads. Two of these pads are used to force the current and the other two are used to measure the voltage. To obtain an average interfacial resistance, the voltage measurements may be made by reversing the current through the original pads and by repeating the measurements with interchanged terminals.

Table 13-3 Backend Testing

Resistor	Capacitor	Junction	Transistor
Contact resistance	Dielectric breakdown and thickness	Leakage	Process-induced charging
Opens/shorts			
Misalignment			Threshold voltage

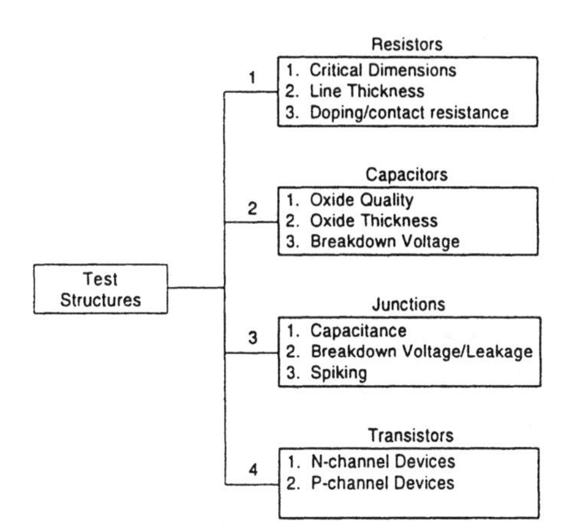

Figure 13-19 Four basic device and process characterization test structures

To measure the effects of topography and processing conditions on contact resistance, contact chains are used. In the case of CMOS devices, these contain several hundred to thousands of contacts over a varying topography, silicided diffusion (n^+/p^+), and polysilicon. The contact and via chains are placed in the scribe streets or in test chips that are strategically located on the wafer. Figure 13.20 shows the top view of a contact chain consisting of contact links. Current is forced through the chain and the resulting voltage drop across the pad is measured. The chains are measured at different locations across the wafer to obtain within wafer variation of resistance. All the wafers in a lot may be tested, and the contact resistance trend by lot provides a very good picture of the lot-to-lot variability and any excursions. The variation of the contact resistance across a wafer may be used to check for systematic processing problems. Contact resistance failures usually have a strong correlation to yield and therefore are primarily used as yield monitors.

A cross-sectional view of these contact links is shown in 13.21. This is a schematic representation of the contact links over diffusion. As shown by the arrows, the current i is forced from one end to the other. The current travels through the metal interconnect and into the plug (P, if used) that connects the diffusion or polysilicon to the interconnect. Due to the layout of this chain, the current is forced to travel through all the plugs, provided the contacts are not open. The number of contacts in a chain is dictated by the Si real estate that can be allowed for test structures. Obviously, with more chains the capture rate of processing defects will be higher.

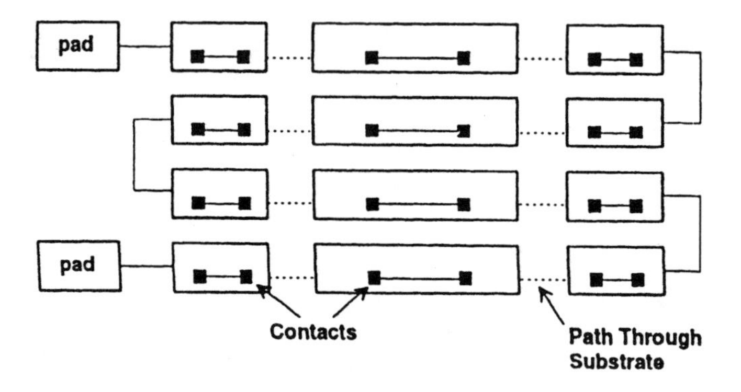

Figure 13-20 A schematic top view of a contact chains test structure for determining contact resistance

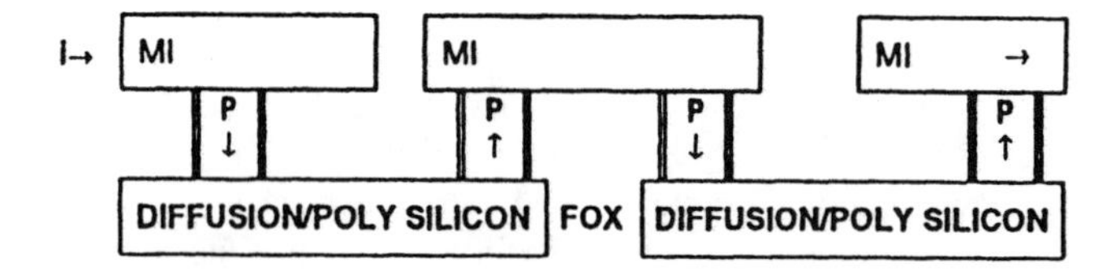

Figure 13-21 A schematic cross-sectional view of contact test structures

Metal shorts/opens. As metallization pitch has decreased, the need to check the integrity between metal interconnects has become very evident. Figure 13.22 shows two generic structures, which may be designed with several variations, for testing for metal shorts or opens at the same level or between levels of interconnects. These shorts or opens may be caused by various processing defects, for example, Al hillocks. To capture the various conditions, both the serpentine and comb structures are defined over a varying topography. The electrical measurement is once again a resistance measurement, and the current is forced through the structure and the voltage drop across it is measured. These structures may be used for electromigration evaluation also.

This concept may be extended to monitoring the defect density of a piece of equipment or a specific segment of the process. Wafers with dice having serpentine and comb patterns may be defined and processed through a small segment of the process that is under evaluation. Testing these structures for opens and shorts will provide a clear picture of the defects that cause shorts or opens and therefore can be traced to particular equipment. Processing such wafers at regular intervals through various segments of the process will help in determining the electrical defect density baseline. The impact of any process changes may be quickly checked by processing such wafers and the data compared with historic values. Furthermore, these wafers may be visually inspected and the defect that caused an open or short characterized.

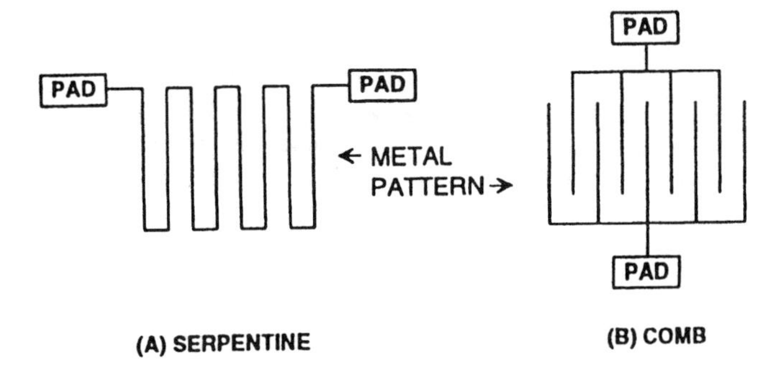

Figure 13-22 Metal serpentine and comb structures for determining shorts and opens in conducting layers

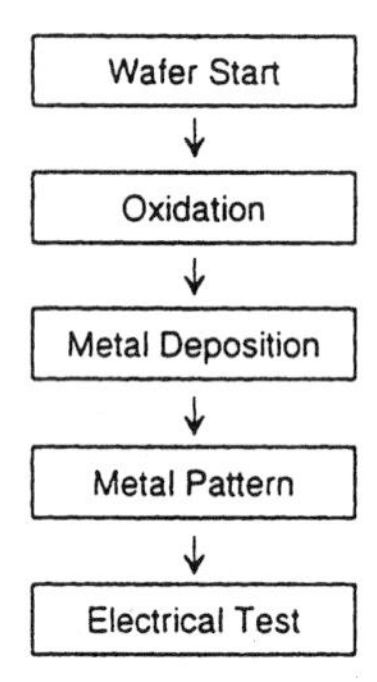

Figure 13-23 A simple process flow for generating defect monitors with serpentine and comb structures

These structures are useful for obtaining information on a host of issues. This information may be obtained within days as the process flow for generating these structures is simple and consists of only a few steps, as shown in Figure 13.23.

Misalignment. The registration between layers, especially critical layers, is an important process control and process development tool. Misalignment may be measured electrically or by using optical means at each lithography step. Electrical test structures can be used to monitor the degree of misalignment and its impact on process problems.

To measure the degree of misalignment between layers using an electrical test structure, a simple resistor with multiple taps for voltage measurement may be designed to detect the misalignment in both the x and y directions. When there is no misalignment the voltage difference between these taps will be zero. With misalignment the resistance changes and the difference between the taps may be measured. This difference gives both the magnitude and direction of the misalignment. This result is an indication of how well the lithography process is controlled. As an aid in verification, optical and visual misalignment structures may be placed next to the electrical structures.

The structure shown in Figure 13.24*a* and *b* and their respective cross sections *c* and *d* are designed to understand the implications of misalignment on the circuit functionality. These structures show misalignment of the metal interconnect (MI) over a contact or via plug, if used. In an ideal situation, the metal would completely cover the plug; in a misaligned state, part of the plug is exposed. Structures with various degrees of misalignment may be designed and placed on the wafer to study the impact of misalignment on contact resistance. For example, contact resistance structures may be defined with misalignment in both the negative and

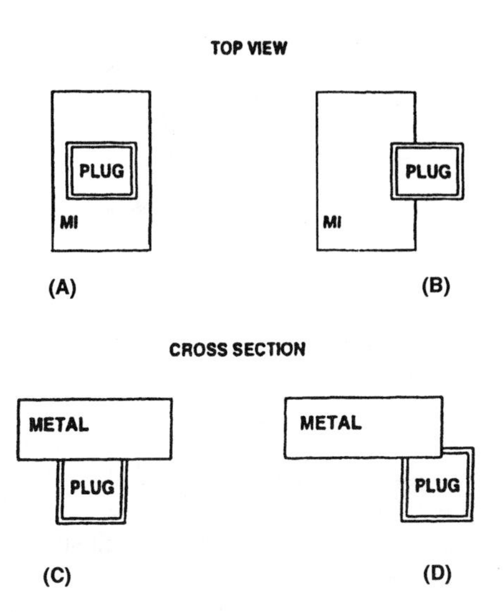

Figure 13-24 Pattern misalignment test structures showing top and cross-sectional views

positive directions with varying magnitudes of misalignment. If contact resistance increases as a result of misalignment, one of the reasons may be plug damage. This damage is a real problem as the plugs would be exposed to the metal etch and cleans. During the metal etch process, the plug may be partially etched to the interface between the metal and the plug may be contaminated. The robustness of any process depends on its ability to tolerate some amount of misalignment. The design rules should allow this as long as there is no detrimental impact on the electrical parameters.

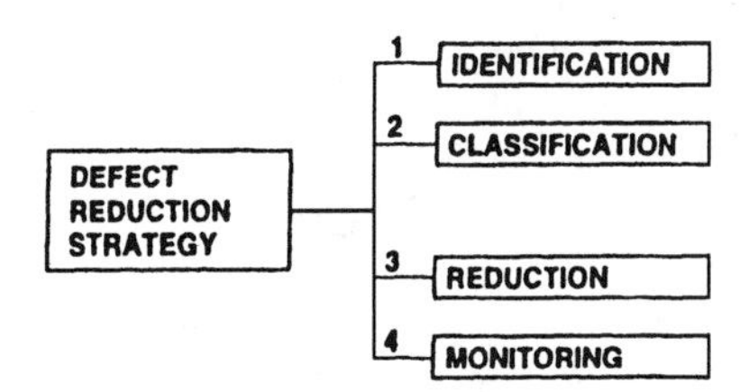

Figure 13-25 Four major areas of focus for defect reduction

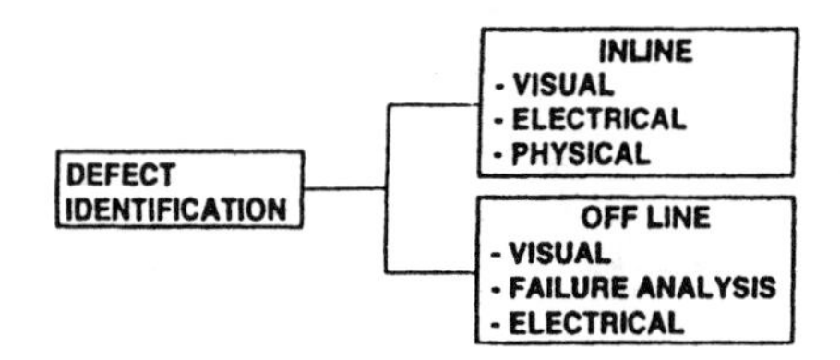

Figure 13-26 Inline and offline defect identification methods

13.5 Defect Reduction

Achieving zero defects, and therefore high yields, is definitely the goal, but the cost of doing this should also be evaluated. Let us discuss this subject by breaking it up into four key defect reduction groups [Donovan (ed.) 1990, Strathman and Lotz 1990, Ramakrishna and Harrigan 1989, Coleman and Chitturi 1990, King et al. 1989, Riga 1977, Soden and Hawkins 1989, Henderson 1989, Liljegren 1992, McCarthy et al. 1988, Bakker 1990, Caludius 1991] as shown in Figure 13.25 and examining each one in the context of multilevel interconnect processing.

13.5.1 Defect identification

Every process will be affected by defects that may depress yields. The key to success is identifying these before significant and severe damage is done to the products. Defects should be identified inline, that is on a real-time basis, as wafers are moving through the production line. As Figure 13.26 shows, defect detection may be done using visual inspections at various locations in the process flow. In addition to this, particle (a loose definition is used here to cover all defects) counters may be used to determine size and location of defects. Lastly, inline electrical testing of serpentine and comb structures provides valuable information on defect levels. If possible, the electrical defect must be correlated to the visual defect in order to understand the failure mechanism.

Offline (end-of-line) defect identification is more elaborate as several failure analysis techniques are used that may not be used in the water fabrication clean room; for example, gold coating of samples for scanning electron microscope (SEM) analysis. Offline defect identification involves isolating the failing bits, rows, and columns using bit mapping and carefully removing layer by layer until the defect that has caused the failure has been found. Once the defect has been found, various visual images are taken. If required, an elemental analysis is done to learn more about the nature of the defect. Obviously, this is a time consuming task. It is important at this juncture to examine the various defect identification methods and their throughput time for analysis. Figure 13.27 compares the various defect

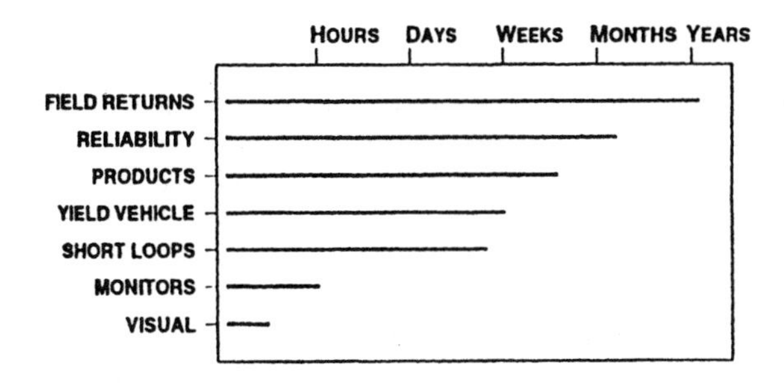

Figure 13-27 Comparison of the delay in information feedback from several defect identification/monitoring methods

identification methods (listed on the y-axis) with the throughput time (x-axis). The throughput times given here are simply for illustration purposes and are not absolute times. The throughput time would depend on the device type, processing time cycle, etc. From this we can conclude that the greatest leverage is in inline defect monitors.

Field returns. As indication of the interaction of a defect and the environment over a long period of time is determined by the failure rate of the devices out in the field. Failure analysis of these failing parts can help in determining the failure mechanism. Feedback to the factory may be late and the defect that caused the failure may or may not still be present. However, the defect identification of field returned parts may shed light on any systematic or random problem with the process. For example, corrosion of interconnects usually gets worse over time and this may cause a failure. If this is a systemic problem, it is then appropriate to go back and revisit the interconnect definition process.

Reliability. For a faster look into the quality of the contacts, interconnects, and dielectrics, reliability tests such as burn-in, steam test, and electromigration are used to determine the failure rate of the parts. The data from such reliability tests take time as devices have to be packaged and the tests performed at different conditions to simulate real-life conditions. These tests can not be used as real-time defect reduction methods, but can provide baseline data for comparison as material is sampled for qualification of a new process or a major process change.

Products. On a more frequent basis, probe yields give a reasonable picture of the level of the defect density over time since yield is exponentially inversely proportional to defect density. To determine the defect density baseline, wafers with a certain yield can be selected at regular intervals and defects determined using offline methods. In this manner, the baseline defect density for each layer can be determined as a function of time.

Yield vehicle. For a more proactive method for defect reduction two methods may be used. The first is to map out electrical defect density and the other visual defect density. In some cases these two correlate and in others they may not. The reason for this is that some defects may not degrade the yield.

Short loops. To obtain the electrical defect density of a section of the process, or just a few steps, various short loops may be used. These can have regular test structures for measuring the resistances, functionality, or just structures that have been designed for measuring the density of opens or shorts. These short loop process flows are powerful aids in establishing the defect density baseline or for the qualification of new equipment at a process step. The throughput time for these short loops may range from a few days to a couple of weeks.

Monitors. For visual defect characterization, SEMs or inline automated defect inspection tools may be used. These tools could be particle counters or more elaborate tools that can count and classify defects through several layers. The feedback is immediate on the type of defects and, if the problem is significant, the production line may be stopped and the defect mode investigated. Inline visual inspection is well positioned to catch excursions and prevent good material from being processed through suspect process steps. The last monitor that helps present a defect density picture is the equipment particle monitor. These monitors highlight any excursions in equipment that might contaminate product.

13.5.2 Defect classification.

The defects identified are classified according to their size and nature. As feature sizes decrease, smaller and smaller defects impact the yield and therefore it is important to classify defects by size. The nature (size, shape, elemental makeup, etc.) is important as it helps focus on a process module that could be the source of the problem. Classification of defects by module and by defect types is essential to get an overall view of the defect scenario. New defects are classified as they are discovered and added to the defect library.

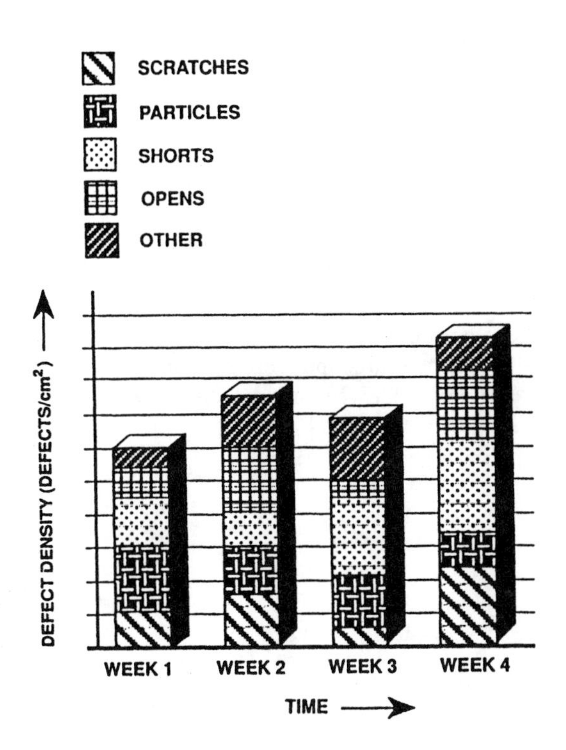

Figure 13-28 A histogram of defect density variation over time

Defect density presentation. Defect paretos are an excellent graphical tool to observe the progress of the defect reduction effort and also to direct valuable resources to addressing the highest leverage defects. In addition to pareto charts, defect status may be presented as shown in Figure 13.28.

In this chart, the defect density for each is presented and gives an overall picture of the defect levels in the factory. The defect density numbers do not pertain to any particular factory and are just used for illustration purposes. The total defect density may be used to determine the yield models. This gives a clear signal for the defect reduction activities on which to focus.

Defect types. The discussion so far has focused on providing an overview of the manufacturing, process integration, yield issues, and techniques for monitoring the defects. This section presents some of the common defects affecting multilevel interconnects that degrade yield and reliability of ICs.

Integrated circuits with multiple levels of metallization have a greater propensity for defects as they entail more processing than an integrated circuit with just a single level of metallization. To aid the reader in understanding defects and how they impact the process and/or product, each defect is discussed as follows:

- Defect source
- Impact on yield and reliability
- Process fix

Such an analysis helps in the evaluation, elimination, and prevention of the defects. The origin of defects may vary but the effect may be classified into two main defect categories. The two categories are as follows: shorts and opens.

Shorts. These are defects that are electrically active and alter the current path when the defects bridge interconnect lines that are physically separated by an insulator. Figure 13.29 shows a typical example of a short between two conducting paths.

Figure 13-29 An example of shorted conducting paths

Figure 13-30 An example of discontinuity (opens) in a conductor

Opens. These are defects that cause a discontinuity in the conduction path. This discontinuity may be missing interconnect material or an interaction with other defects that might have prevented the formation of the interconnect path. Figure 13.30 shows opens in interconnect lines due to defects.

The following discussion considers a few typical defects that affect multilevel interconnect processing. These are divided into two categories: systematic and random. It is not always easy to classify the defects in such a manner. Defects may switch from one category to the other easily depending on the nature of the defect source, which may be the process or equipment. Table 13-4 highlights some of the common defects.

Table 13-4 Common Defects

Systematic	Random
Corrosion	Particles
Hillocks	Mask defects
Edge effects	Scratches

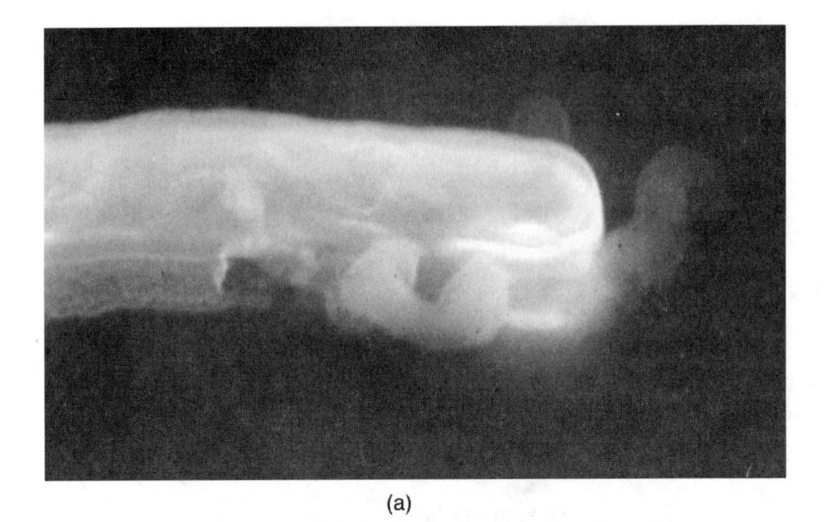

(a)

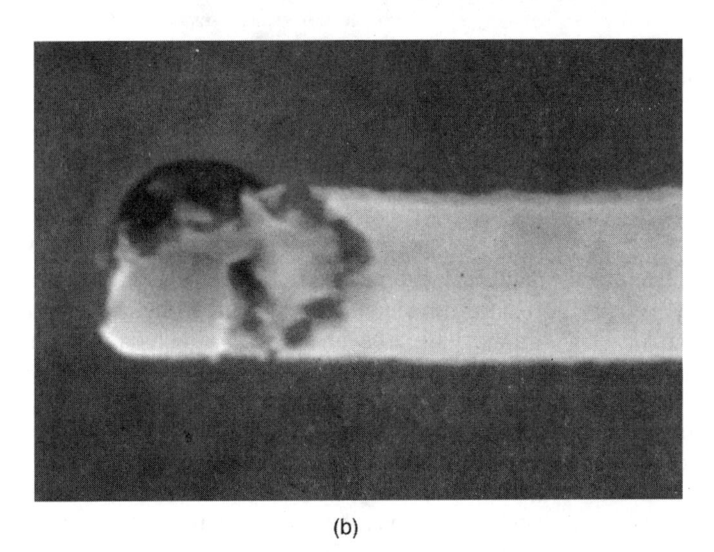

(b)

Figure 13-31 Corrosion in metal interconnects. *(a)* Creating a potential short. *(b)* Causing an open

Let us take an example of this dilemma in defect classification. If a particle were to create a defect on the mask, then this defect occurrence may be considered as random in nature. However, if this mask is then used to print the pattern on the wafer, we would have a systematic failing location owing to the step and repeat operation of the stepper.

Corrosion

Source. The primary source of corrosion is the metal etch process chemistry. Aluminum interconnects are etched using chlorine (Cl) which is readily absorbed in both resist and the oxide surface. Corrosion is observed if the wafers are exposed to atmosphere following etch without passivation or some chlorine removal step beforehand.

Impact. If corrosion is widespread, it may be possible to detect it on wafers while they are still in the clean room. Sometimes, corrosion of metal interconnects takes a long time and could cause a failure after the parts have been packaged and shipped. Corrosion may either create shorts or opens. Shorts are created when the corrosion byproducts connect two adjacent interconnects. In some cases, corrosion removes part of the interconnect material leaving a discontinuous or open interconnect. Figure 13.31 shows the possible ways corrosion can cause device failure. Figure 13.31*a* causes an interconnect short and Figure 13.31*b* causes an interconnect open.

The impact of corrosion may be observed in several ways depending on the severity of the problem. Parametric testing after wafers have been completely processed may show a change in metal interconnect critical dimensions. In addition, shorts or opens may be observed between interconnects. These problems translate into yield loss if the corrosion reaches an advanced stage.

Fix. Corrosion problems associated with metal etch and cleans may be solved by an effective in situ passivation step that is a part of the etch process and is followed by an efficient resist cleaning process. Immediate substitution of Cl by fluorine (f) absorbed in the resist followed by resist stripping is done to prevent corrosion. Capping the Al interconnect with a dielectric seals the film and prevents further corrosion.

Al hillocks

Source. One of the common problems with Al interconnects is the formation of hillocks. *Hillocks* are extrusions of Al from the surface of the film that are caused by the compressive stress in the metal film. The problem is severe for Al interconnects because of the thermal mismatch between Al and Si, especially since Al adheres very well to Si and SiO_2. When the Si wafer is heated, the different rates of thermal expansion cause the Al film to expand more than Si or SiO_2. This resulting stress causes formation of hillocks. Compressive stresses in the Al film become high as the temperature exceeds 300°C, a very easy temperature limit to exceed in most chemical vapor deposition (CVD) dielectric film deposition processes. In addition to this mechanism, hillocks may also be formed by the high rate vacancy diffusion in the Al films. Figure 13.32 shows a typical Al hillock that has been formed after thermal treatment.

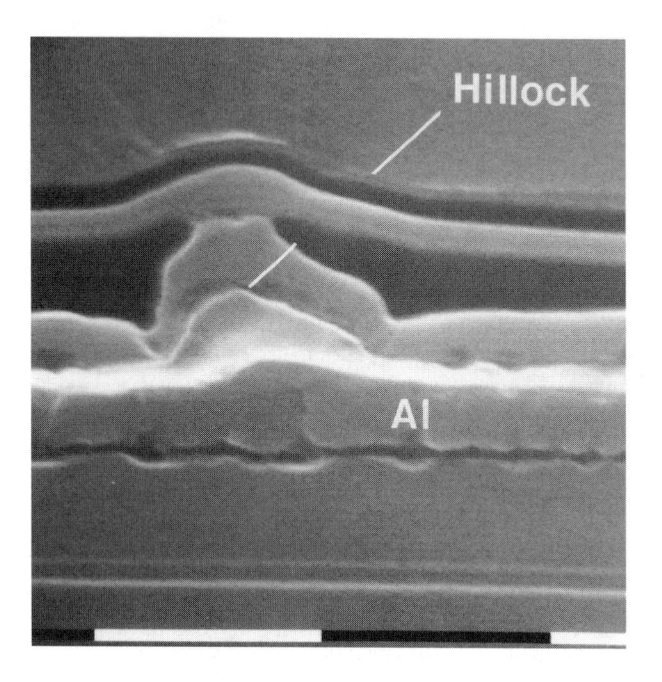

Figure 13-32 A hillock in an aluminum film

Impact. The impact of hillocks on interconnects is detrimental to the device functionality. Hillocks cause shorts in both the lateral and vertical directions. In the lateral direction, two adjacent metal lines may be shorted; a problem that gets worse as the metal pitched are reduced. In the vertical direction, the hillock is covered by dielectric. In subsequent via definition steps the resist over the hillock will be thinner than in other locations. During the etch step, the resist erodes quickly, exposing the dielectric to the etch, which may completely remove it. The hillock could short to the next level of metallization.

Fix. There is no one comprehensive fix for the hillock formation problem. There are, however, several techniques [Wolf and Tauber 1986] that help reduce hillock formation. A few of the key ones are given here. Any fix

Figure 13-33 Wafer edge damage during processing, highlighting a potential source for defect generation

to the hillock problem must be considered along with the other integration issues associated with changing the composition of the metal stack.

- Restricting the dielectric deposition temperature to around 350°C. In addition, the hillock size and density may also be minimized if a low compressive stress dielectric film is deposited over the metal quickly enough.
- Capping the Al film with a layer W or Ti.
- Adding Cu to the Al film. This has been shown to minimize hillock formation as the Cu precipitates in the grain boundaries and prevents vacancy migration.

Edge Effects

Source. The wafer edge is a source of defects which may not stay at the edge but may cause shorts/opens anywhere on the wafer. These defects may be transported and deposited on the dice as the wafers go through various processing steps. Figure 13.33 shows a wafer edge that has been subjected to the extremes of various processes. Particles and other defects around the edge could easily cause problems. Figure 13.34 shows the area on a wafer that is susceptible to damage.

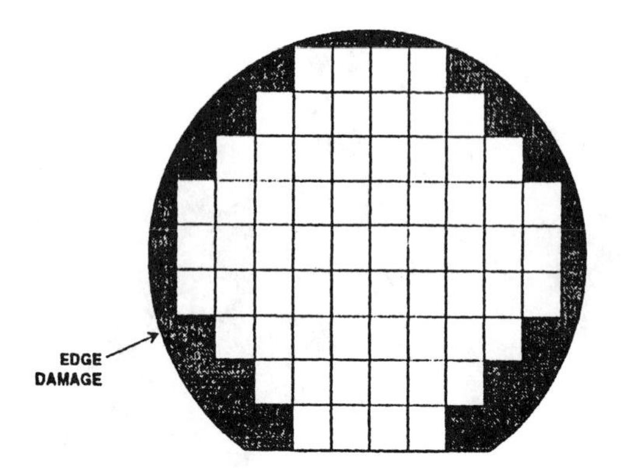

Figure 13-34 A wafer map indicating areas of damage around edge

One of the sources of this damage is the placement and interaction of the various machine clamps that are used to hold the wafer in place. Damage can also occur when resist is removed from the wafer edge as part of the photolithography process. This edge bead removal process, as it is called, may intrude in from the wafer edge. Since dice are not printed in the shaded area, microloading effects during the etch process further damage this area due to no pattern and inadequate film thickness, arising from film non-uniformities at the edge.

Impact. The impact of not controlling and monitoring edge effects can be severe as edge dice adjacent to the shaded area may not be reliable due to defects from the wafer edge. The short is the common failure mode associated with the edge contamination. This translates into yield loss not only on the edges but sometimes in other locations.

Fix. Can this shaded area be reclaimed and die printed there? Perhaps some amount of Si real estate may be recovered with optimization of the deposition, etch equipment setup, and other processes. We cannot help but lose a fraction of the area around the wafer edge. The short-term fix in most cases is to compensate for aggressive etches around the edges with a thicker oxide film. The long-term fix is a combination of equipment and process modification. One other technique is to print all the dice to the very edge.

Particles

Source. Particles may or may not cause yield loss, but it is a good policy to reduce the particles in the production area and on the wafers. Figure 13.35 is a cause and effect diagram for particle generation that is essentially an expansion of the branch given in Figure 13.16. Particles may be deposited on the wafer surface due to the mechanical movement of wafers through equipment process chambers. The wafer transport mechanisms in most machines are a major source of particle generation. What about process induced particles? This is the other main component to the problem of particulate contamination. Contamination can be generated from any process steps. For example, process-induced particulate contamination in dielectric or metal etch chambers is a major concern. The etch process relies on polymer formation to control profiles. This polymer gets deposited on chamber walls and flakes onto wafers. Figure 13.36 shows a particle that has been deposited from chamber walls in a deposition system.

Impact. Particles may be benign at the layer they were deposited, but may cause catastrophic yield loss at some downstream location. This occurs when the particle serves as a nucleation site for downstream depositions of metals and has a layer of metal deposited over it. After the etch process, this particle could have a metal spacer around it that could short interconnect lines. This is one mechanism where small particles increase in size as they are exposed to downstream processes. Particles degrade yield by causing shorts, opens, or both.

Fix. Obviously, it is of paramount importance to reduce the particulate contamination levels. One way is to improve the equipment that is used and the other is to develop processes that operate within a window of very low particulate contamination. Equipment-related particle generation is one area for continuous improvement to realize defect density goals for each part or module of the process flow. Control and reduction of particles in process equipment are all the more critical when the same set of machines is used several times in the process. For example, metal deposition machines are used several times for interconnect definition as described above. In some cases, reduction in particles is only possible through the use of wet cleaning procedures. Unfortunately, wet cleaning options are limited for interconnect formation steps. Wet cleans may be divided into two modes of application.

- *Film removal.* This is usually removal of photoresist and some amount of dielectric. Incomplete removal of these films may cause catastrophic failures. For example, after metal etch, the photoresist must be removed and the Cl absorbed surface cleaned to prevent metal corrosion.

- *Contamination removal.* This is critical for contact and via layers where residues, particles, and other contamination may prevent good contact and cause the resistance values to increase. Cleans used here must be compatible with the metal layers and environmental requirements.

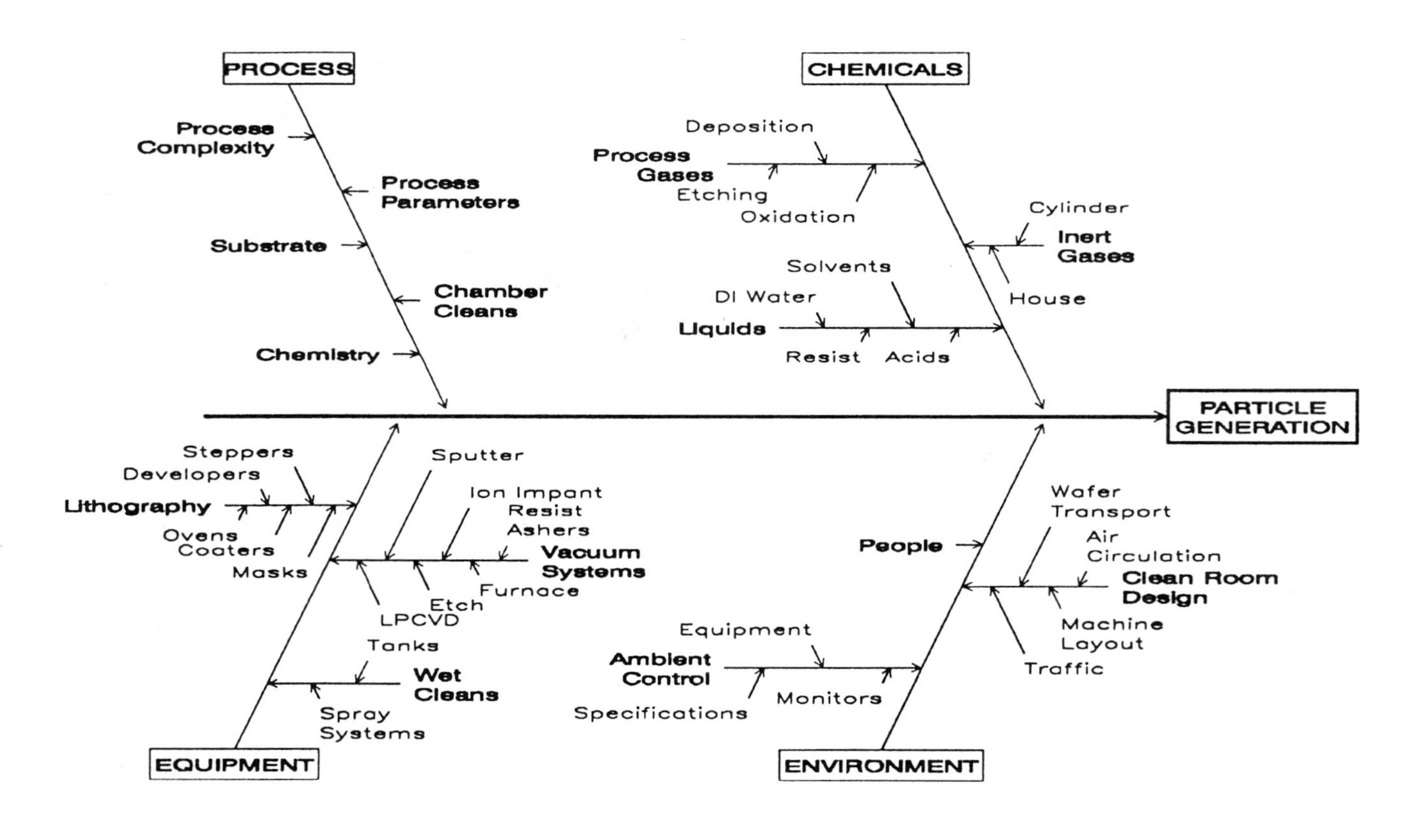

Figure 13-35 Cause-and-effect diagram for particle generation

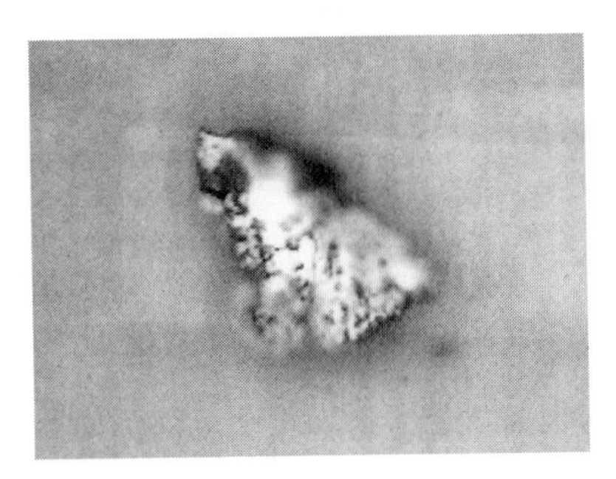

Figure 13-36 An example of particle generation from thin-film deposition systems

Mask defects

Source. A reticle or mask defect may be random in nature and might have been caused by particles or by residues from the mask-making operations. However, this defect becomes systematic in nature after the reticle is used to print die on Si. Let us assume that the reticle field consists of two dice that are stepped across the wafer. If one of these has a defect, then the defect is also printed on the wafer as the stepper moves across the wafer. Figure 13.37 shows the situation where we have a repeating bad die that alternates with a good die.

Impact. The impact of such a defect is catastrophic and results in a 100 percent yield loss, assuming that the reticle field consists of two dice. The defect could cause shorts or opens depending on its nature and placement.

Fix. The fix for this problem must start in the mask-making operation. Stringent quality control of the product (reticles) is required. Once these reticles arrive at the fabrication site, careful inspection and sample printing across the wafer are often done to ensure that no pattern defects exist.

ILD pinholes

Source. The primary source of interlevel dielectric (ILD) pinholes is the thinning and subsequent stripping of resist over the tallest parts of the die during contact/via etching. Aluminum hillocks and particles also cause pinholes in the dielectric film.

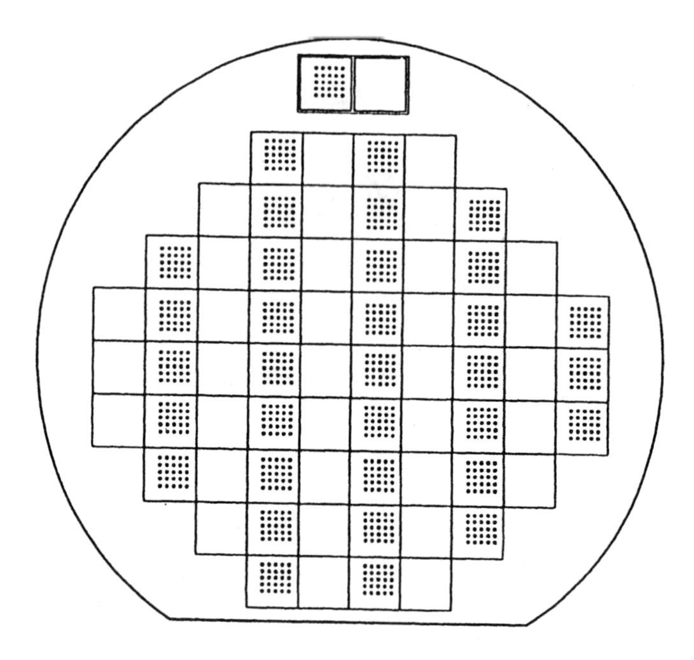

Figure 13-37 Yield loss due to repetitive mask defects

Impact. Location of the pinhole determines the impact on circuit functionality. If the pinhole were to fall over an area where two conductors overlap, then these would be shorted, at there is now a path for conductors between them. In such a case the yield is degraded. Reliability of the die will be degraded if the dielectric has pinholes. These pinholes can trap moisture and other contaminants. This is especially true for the passivation layer which is deposited to provide a hermetic barrier and protect the circuitry underneath. Pinholes in the nitride passivation layer can cause reliability failures. A quick check for pinhole density in the dielectric film is to dip the wafer in an acidic solution. If pinholes are present, then the acid will attack the Al film underneath as it penetrates through the pinholes.

Fix. Pinholes may be prevented by reducing the particle density and hillock formation. Increasing resist thickness and improving the oxide to resist selectivity are two of the techniques that may prevent pinholes from forming. Of course, the thin oxide film that is deposited must be of good quality to begin with. In some cases multiple depositions are done to reduce pinhole occurrence.

Scratches

Source. The source of scratches is primarily wafer handling within equipment itself or by people handling wafers. Scratches may be a random defect but this is really a systematic problem that impacts all processes.

Impact. Scratches may rip the interconnect lines and cause opens. The scratches will be areas of high defect density as the wafer moves through the process. There is always a danger that defects from these scratches may contaminate neighboring dice.

Fix. No or minimal wafer handling by manufacturing operators is the ultimate solution. Equipment-induced scratches may be addressed with modifications to the wafer handling system. Automated cassette to cassette inspection and monitor systems prevent the need to handle wafers. Automated systems for wafer transfer are used to minimize handling.

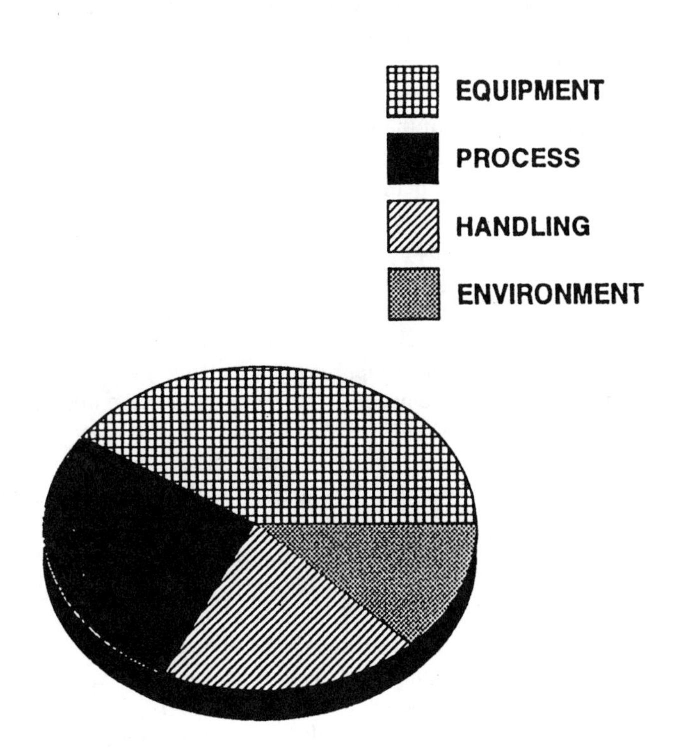

Figure 13-38 Sources of defects

13.5.3 Defect reduction

Having identified and classified the defects, the next major program is the elimination of these defects. From the yield loss and particle cause and effect diagrams, we observed that there are several sources of defects. The defect contribution varies from factory to factory, quality of equipment, process, and raw materials. Figure 13.38 shows a pie chart with a possible breakdown of the defect contributors to overall defect levels. Such an illustration is useful in launching a factory-wide campaign in defect reduction, involving vendors, improving incoming quality of materials, proper selection and evaluation of process equipment, ultimately enhancing the manufacturing methodology. If these basics are not addressed, defect reduction becomes that much more difficult. Let us examine the defect reduction thrusts for the equipment and process more closely, as shown in Figure 13.39.

Equipment. There is no question that the quality and design of equipment is extremely important in reducing defect levels. For example, a wafer movement mechanism based on belts and exposure to high temperature tends to have higher levels of particles than a "pick and place" type of system. Improvements in equipment must be made by the equipment vendors to satisfy technology demands of the future.

Equipment downtime should be monitored and the causes addressed. Obviously, downtime creates manufacturing bottlenecks. The solution is not increasing the capacity by purchasing more machines, but enhancing the capacity by increasing the mean time to failure of the equipment. This is done by timely and thorough preventative maintenance (PM), even in the face of production pressures. Some processes cause polymer or other chamber coatings on the equipment chamber walls. Flakes from the chamber walls may fall on the wafer surface, causing yield loss. Chamber cleans, either in situ of not, are needed to prevent an increase in defect levels. Equipment particle levels should be monitored on a regular basis using particle counters. Excursions in particle levels should be reason enough to shut down the machine, investigate the cause, and take corrective actions.

Process. Defects that become killer defects as a result of process interactions may be effectively eliminated by defect partitioning. Defect partitioning, a defect identification technique, involves inspecting the same area on the same wafer as it moves through the process flow. The results are studied by overlaying the location of defects at each step. This provides information on how each process changes the nature of the defect and contributes new defects. The success of defect partitioning depends on the accuracy of inspection tools in locating defects and their coordinates. Pictures may be taken after each step so the changing appearance of the defects may be observed. Using detailed pictures and elementary analysis, a

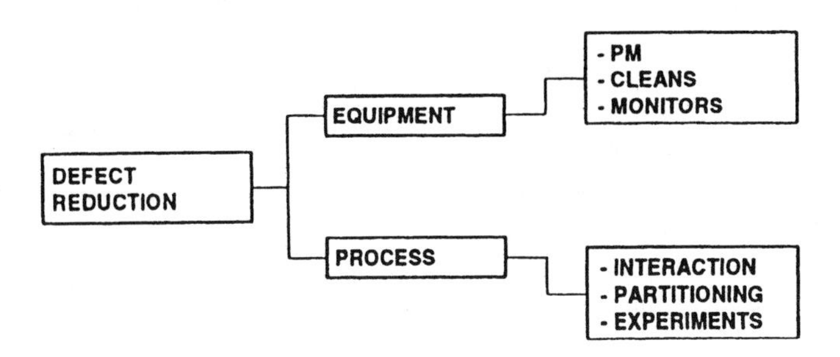

Figure 13-39 Defect reduction directions

cause and effect diagram should be generated to highlight all possible causes of the defect(s). The various models that are generated by these problem-solving exercises may be evaluated by split-lot experimentation. To speed up resolution of problems, people who have a broad and good understanding of the process should be consulted.

13.5.4 Defect monitoring

Defect monitoring procedures must be carefully designed and implemented since they tell us if we have been successful in defect reduction or not. This step may be divided into two activities, as shown in Figure 13.40. Earlier we discussed the techniques for establishing the baseline defect level for a process. The baseline defect level is presented either in trend or pareto charts and appropriate action is taken to continuously reduce the baseline. Once a defect excursion had been isolated and a fix instituted, it must be monitored to make sure that the defect is truly eliminated. Monitors for this purpose may be manual or automated inspection. Proper documentation is necessary to finally complete the whole problem-solving process.

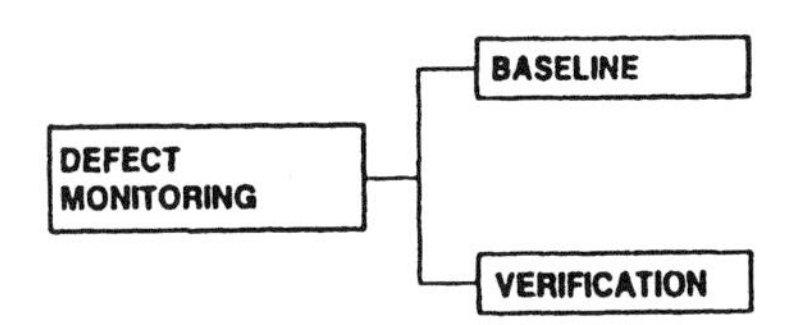

Figure 13-40 Defect monitoring procedures

13.6 Summary

The focus of this chapter has been on manufacturing issues – merging the designs and processes to obtain high-yielding wafers. Yield enhancement is possible only when proper manufacturing practices are combined with defect reduction activities and cost boundary conditions are satisfied. Truly, the rate-limiting step in producing ICs in large volume ahead of the competition lies in the successful understanding of manufacturing technology. As process technology is pushing the level of integration to higher and higher levels, the need for more levels of metallization increases. Such an increase in interconnect levels leads to several cycle-time improvements and defect reduction opportunities.

14. QUALIFICATION

Qualification programs include a set of tests that are performed in order to transition a semiconductor product or product family from the engineering phase into the manufacturing phase [Pecht 1994]. Qualification is also conducted when the manufacturer considers a change made to the product material or manufacturing process 'significant'.

Once the part has passed qualification testing, periodic reliability monitoring tests are used to prevent products made with substandard, materials or manufacturing processes from being sold. Periodic reliability monitoring tests are also used to alert the manufacturer if process controls fluctuate beyond acceptable levels. In general, the test conditions used for periodic reliability tests are less stringent than those used for qualification tests.

14.1 Qualification and Reliability Monitoring Testing

Common tests include:

- High Temperature Operating Life (HTOL): This test is conducted to precipitate diffusion and drift based failure mechanisms. Devices are exercised at elevated temperature (125 - 150 °C) and voltage (Vdd - 1.5 Vdd), and at 1 MHz or higher frequency, for 1000 to 2000 hours. The test population is taken off the life test and is tested at a number of read-out points, typically covering 168 hr, 512 hr, 1008 hr intervals. With the modern high pin count units, the devices need to be exercised by an intelligent set of test vectors, rather than random or pseudo-random patterns, thereby adding considerable complexity to the normal life test systems.

- Autoclave (Pressure Pot): This test is conducted to determine the moisture resistance of non-hermetic packaged devices. No bias is applied to the devices during this test. Autoclave employs severe conditions of pressure, humidity, and temperature not typical of actual operating environments, that accelerate the penetration of moisture through the external protective material (encapsulate or seal) or along the interface between the external protective material and the metallic conductors passing through it. When moisture reaches the surface of the die, reactive agents cause leakage paths on the surface of the die and/or corrode the die metallization, affecting DC parameters and eventually catastrophic failure. Other die-related failure mechanisms are activated by this method including mobile ionic contamination and various temperature and moisture related phenomena. The test is destructive and produces increasing failure rates when repetitively applied. Autoclave tests can also produce spurious failures not representative of device reliability, due to excessive chamber contaminants. This condition is usually evidenced by severe external package degradation, including corroded device terminals/leads or the formation of conducting matter between the terminals, or both. Specific device types and technologies may be particularly sensitive to package degradation. Steam is heated to 121°C and 15 psig. The saturation conditions are maintained uninterrupted for the period of time specified in the test procedure, usually 48, 196, 240, or 500 hours [QSI, 1996].

- Temperature Cycle: This test is conducted to determine the resistance of devices to alternate exposures at extremes of high and low temperatures. Permanent changes in electrical characteristics and physical damage produced during temperature cycling result principally from mechanical stress caused by thermal expansion and contraction. Effects of temperature cycling include cracking and delamination of packages, cracking or cratering of die, cracking of passivation, delamination of metallization, and various other changes in electrical characteristics resulting from thermo-mechanically induced damage. Devices under test are placed in a test chamber with hot and cold sections in such a way that the airflow around each device is unobstructed. The devices are then cycled between the hot (+125, +150°C) and cold (-55, -65°C) chambers, with essentially "zero" dwell time at room temperature, for the number of cycles (500, 100) specified in the detailed test procedure [QSI 1996].

- Highly Accelerated Stress Testing (HAST): These tests are conducted to evaluate the reliability of non-hermetic packaged devices operating in humid environments. HAST testing is a steady-state temperature-humidity-bias life test. It employs severe conditions of temperature, humidity, and bias which accelerate the penetration of moisture through the external protective material (molding compound, or encapsulant) or along the interface between the external protective materials and metallic conductors passing through it. When the moisture reaches the surface of the die, the applied potential forms an electrolytic cell, which corrodes the aluminum, affecting DC parameters through its conduction, and eventually causes catastrophic failure by opening the metal. Typical environments are set at 130°C and 85% RH. The devices are continuously energized throughout the duration of the test. At the end of the specified stress duration, usually 200, 500, 100 hours, the devices are tested [QSI, 1996].

- Mechanical Shock: This test is conducted to determine the suitability of the devices for use in electronic equipment which may be subjected to moderately severe shocks as a result of suddenly applied forces or abrupt changes in motion produced by rough handling, transportation, or field operation. The devices are subjected to shock pulses typically with a peak acceleration 1,500 times the force of gravity (1,500 G), for at least

0.5 milliseconds. Each device is generally shocked several times in each direction on each of its three axes.

- Vibration: This test is conducted to determine the effect on devices due to vibration in a specified frequency range. Vibration frequency is usually varied approximately logarithmically between 20 and 2000 Hz and is traversed for not less than a few minutes. This cycle is typically performed four times in each orientation, X, Y, and Z.

An empirical, product-by-product approach to qualification is expensive in terms of dollar costs, human resources, and product schedules. Typically, a complete set of life tests required to demonstrate the integrity of a new design requires hundreds of test units, and thousands of hours of life tests. These associated expenses can add up to cash costs on the order of tens and thousands of dollars. In addition to the dollar expense, qualification test development requires inter and intra-department coordination. Performing the life tests, building the life test systems, and conducting failure analysis call for a unique blend of engineering skills. In addition to this human resource expense, qualification tests typically require three to six months to complete.

Applying conventional methods to modern deep sub-micron components is becoming more difficult. Performing physically meaningful HTOL tests - the 'staple' of all conventional qualifications - on increasingly complex ICs, with sufficient 'toggle coverage' (% of nodes that are wiggled during the life test) at a frequency comparable to in-use conditions, and at elevated temperature and voltage on 10s or 100s of parts while managing noise, power dissipation, ESD and latchup problems, adds up to a significant challenge. In addition, the interpretation of these life tests is increasingly complex. The traditional Arrhenius type 'acceleration factor' required to relate the life test to in-use conditions is becoming less credible [Pecht, 1997]. Further, life test samples as small as 20 to 30 parts drawn from one or two production runs are not unusual due to cost and schedule constraints; these tests, however, yield results that are statistically meaningless, and incapable of identifying problems associated with process variability (usually the cause of any reliability issues).

The industry has consequently evolved qualification procedures from the conventional product-by-product approach to the "generic" or family qualification approach [Radojcic 1986, Pecht 1994]. In fact, with deep submicron ICs and the associated trends towards "system-on-a-chip", which necessarily raise the issue of importing IP (intellectual property) blocks from ASIC vendors or third party suppliers, implementation of a more advanced qualification methodology becomes imperative [Radojcic 1996]. Both the laws of physics (PoF based reliability management and predictions) and the laws of economics (much reduced time-to-market and product life extensions) favor the evolution of a new set of qualification standards.

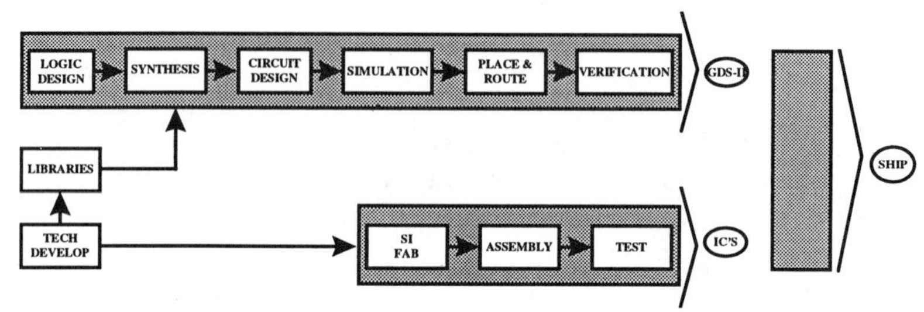

Figure 14-1 The typical serial nature of the conventional qualification

A physics-based approach for managing the chip feature counts (such as number of transistors, total area of gate oxide, etc.), such as the "Universal Qual" approach [Radojcic 1992], must evolve. Figures 14.1 and 14.2 show this potential evolution.

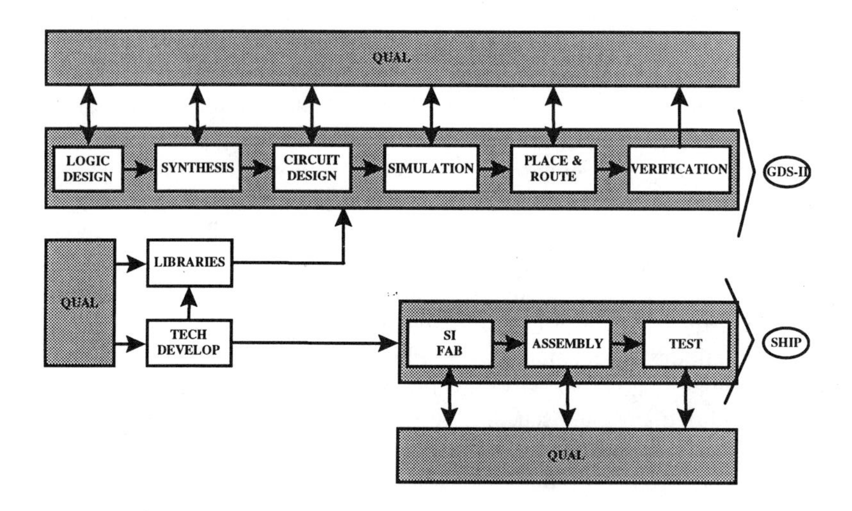

Figure 14-2 The parallel nature of advanced qualification being met

With the product-by-product approach, where a product is individually qualified, little attempt is made to extrapolate data from one design to the next. With the generic qualification approach, where the entire product family and process technology is qualified via one, or few, suitable test vehicles, reliability estimates for individual products are derived through procedures that ensure compliance of that product with the qualified "envelope." That is, worst-case failure rate estimates for a technology and/or product family are generated empirically, and individual products are qualified purely on the basis of similarity to the test vehicle. With more advanced qualification methodologies, the individual physical attributes (rather than individual products or the elements in an ASIC cell library) of a technology are qualified and product reliability estimates are "synthesized" through a set of models based on product complexity.

14.2 Virtual Qualification

To meet cost and schedule objectives, reliability rules are best verified through design simulation rather than through actual qualification testing. A new approach, virtual qualification, is being created. Its flexibility will ensure the quality and reliability of the fielded system in a cost effective and timely manner. The goals of this new approach are to assess the risk posed by an electronic part and to recommend corrective action for improving its reliability. Benefits of this approach include:

- Developing more complex products with more features and higher quality at lower cost,
- Providing shorter design cycles and faster times to market, and
- Providing concurrent design for quality, reliability, manufacturability, and testability.

The strategy begins in the early stages of product design. The strategy is based on a detailed understanding of the design, physical layout, materials, material interfaces, assembly processes and potential failure mechanisms (physics-of-failure concept) to which the product is sensitive by virtue of its life cycle conditions. Market or customer requirements are determined, and an analysis of design and manufacturing capabilities is obtained. As a result, information management and engineering analysis form an essential part of the product development process.

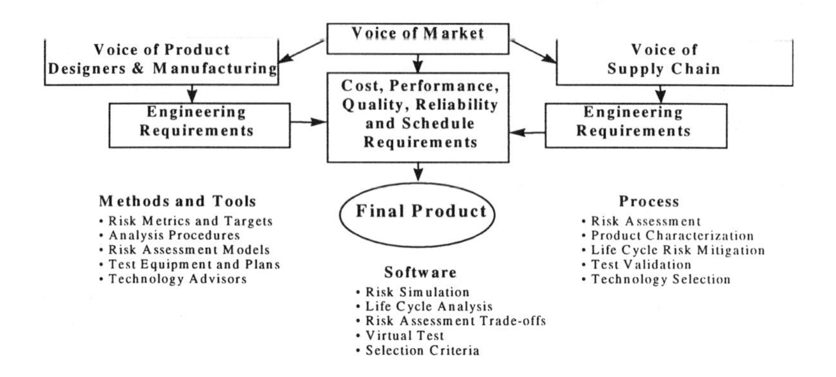

Figure 14-3 Virtual Product Development & Qualification Process

A decision support tool is also embedded into the process that is used to assess and mitigate the risks with physical (size, weight, substrate routing), electrical (delays, attenuation), thermal (internal and external thermal resistance), reliability (lifetime), and cost attributes. An overview of the integrated system is illustrated in Figure 14-3.

An approach for effective risk management through virtual qualification begins with IC and package definitions using the package design interface depicted in Figure 14-4, followed by stress analysis. Because there is often a dependence of component and product failures on operating temperature extremes, temperature cycle magnitude, or temperature gradients, thermal analysis is integral to designing-for-reliability. This is particularly true for high power systems where power dissipation can considerably increase the operating temperature above ambient.

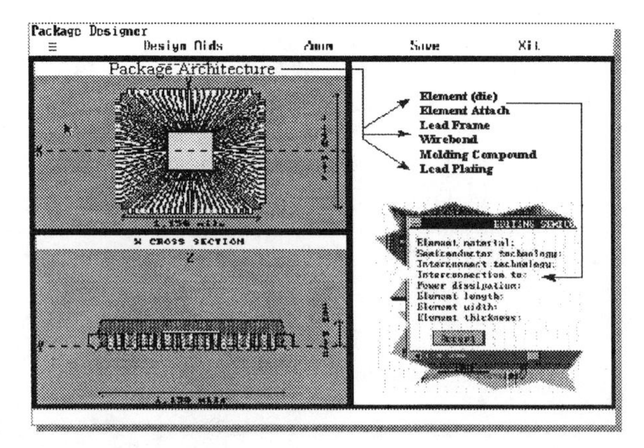

Figure 14-4 The graphical package design interface for PBGA

After the stress analysis is complete, the reliability assessment software is used in any one of four modes. The first mode is virtual qualification of an existing component design. In this mode, dominant failure mechanisms are ranked, starting with those that cause failure in the shortest amount of time. This allows the user to determine the dominant failure mechanisms and the time-to-failure for the package.

The second mode is a what-if study in design-for-reliability. Here, the reliability effects of variations in design attributes and in environmental and operational stresses are evaluated. An example of how software may be useful when conducting a tradeoff between a design parameter and a stress, in order to maintain the same design life for a MCM-L package, is shown in Figure 14-5. Results are used to reduce the risk of package failure by the identified failure mechanisms.

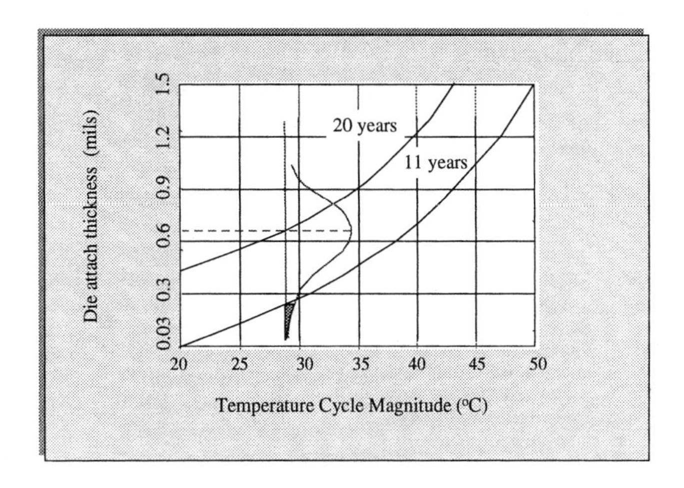

Figure 14-5 Uniform time-to-failure plots for die attach fatigue in an MCM-L package with failure probability distribution. The graph shows that for a mean value of die attach thickness of 0.65 mils 50% of the parts subjected to a temperature cycle of 28°C will fail by 20 years, however, a much smaller fraction will fail earlier, after 11 years. This smaller fraction is those samples in the distribution having a die attach thickness less than 0.25 mils, as represented by the shaded portion of the probability distribution.

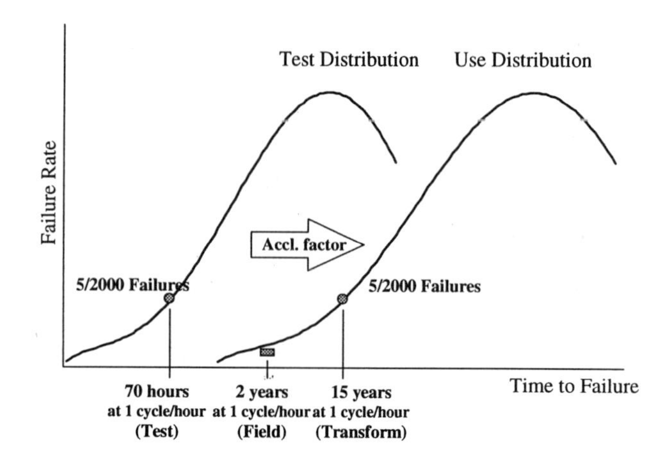

Figure 14-6 Use of acceleration transforms to project field failure distribution. Field failure rate of 5/2000 will be observed at 15 years of operating life if the same failure rate is observed at 70 hours in accelerated test conditions. The acceleration factor is determined through the use of what-if analysis in the software.

In the third mode, acceleration factors must be used to transform a failure distribution at accelerated test conditions to a projected failure distribution at standard use or storage conditions. This is illustrated in Figure 14-6, which shows the failure distribution in test, the projected field failure distribution, and the validation with actual field failure data. In this modem qualification test parameters useful for each package design and general environmental application(s) are selected.

In the fourth mode, the effect of manufacturing tolerances on failures, such as the effect of variations in die attach thickness on subsequent die attach fatigue failure must be evaluated, as shown previously in Figure 14-5 . The time-to-failure for each value of die attach thickness and each temperature cycle magnitude can be identified by their intersection on a line of uniform time-to-failure. The superimposed tolerance distribution shows the percentage of samples having each value for die attach thickness. Together these graphs can be used to determine the percentage of samples which will survive for a given lifetime based on this mean value and tolerance for each die attach.

The software package CADMP-II is a reliability assessment tool that

- Allows components and systems to be qualified based on an analysis of the susceptibility of their designs to failure by a number of fundamental physical and chemical mechanisms.

- Assesses candidate and existing package or module designs for reliability in different environments using physics-of-failure models.

- Calculates times-to-failure for the fundamental mechanisms that cause IC, hybrid and multichip module package failure.

- Evaluates the effects of various manufacturing processes on reliability by calculating time-to-failure as a function of manufacturing tolerances and defects.

- Facilitates the selection of efficient and cost-effective test parameters for validating the reliability assessment, thereby assisting in the design of value-added qualification processes and tests; and

- Facilitates the selection of high reliability components for unique applications by permitting their virtual qualification.

The software can also be used to evaluate the effect of manufacturing defects, such as the effect of die attach void size on failure by the same mechanism.

Testing products to assess field reliability inhibits rapid development of cost-effective electronics. Testing offers no return-on-investment. Testing requires equipment, person hours, test plans including multiple stress condition sequences, products to be put on test, and most significantly time and money. This is especially true for high reliability electronics, where the number of test products and the required test hours can be of the order of the total product development life. In particular, Pecht's Law shows that the accelerated test duration for qualification will have to be five times higher every three years (see Figure 14-7).

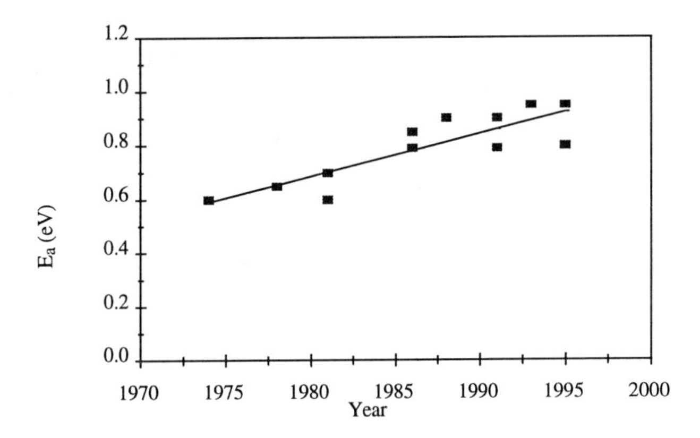

Figure 14.7 Increase of activation energy for device failure mechanisms over the years shown here validates Pecht's law that the semiconductor device acceleration factor increases fivefold every three years. A corollary to this law states that "The accelerated stress levels or test duration (or some combination thereof) for device qualification will have to be five times higher every three years."

15. SCREENING

Product quality and reliability are dependent upon the screens[1] used to detect defects that are introduced during the manufacturing process. These defects may originate from the manufacturing process itself, or from the materials used for the part or those used in the making of the part. The focus of this section is on the specific tests used by manufacturers to screen out defective products.

A 3-sigma process, with +/-3 sigma process distribution fitting within the specification limits, is a well-known "threshold of excellence" that ensures 99.73% yield [Mikel 1990] to that specification. This process results in approximately 2700 failures per one million parts, which is generally unacceptable. Effective screens must detect those 2700 failures. A higher sigma process, e.g., a 6-sigma process, reduces the number of failures to 0.002 failures per one million parts, at which point screens may be considered uneconomical and may be eliminated.

15.1 Functional Tests

Functional tests are used to determine if a product is able to perform to its electrical specifications. Functional testing simulates operating conditions through the application of either digital patterns or parametric analog stimuli to device input pins; output pin signals are monitored. During digital testing, input signals can be a logical high or a logical low. For analog testing, in-circuit tests are often performed, in addition to parametric tests at the chip I/O. Tests are often performed at elevated voltage or temperature values to ensure that there are no marginal defects.

With complex ICs the number of functional test vectors required to fully test the component, and the effort required to produce all the vector combinations, are excessive. Consequently, for digital products Automated Test Pattern Generators (ATPG) are used. ATPGs algorithmically produce a set of vectors required to test all the stuck-at failure modes [Bateson 1985, Hnatek 1995, Jones 1989]. The ATPG methodology, as well as the more advanced Built-In Self Test (BIST) methodology, requires that the test considerations are included in the design of the chip.

The effectiveness of functional tests is determined through fault coverage for single stuck-at faults; if every possible fault that results in a "stuck-at" zero is observable on the output pins, then the fault coverage is said to be 100%. If a product screen results in relatively low fault coverage, then there is a high probability that a defective product will be shipped. With complex die, and/or with processes having high defect densities, the probability of an "escaped" defective part is high and fault coverage on the order of 99% is required. These escaped defects can cause large fall out at the system integration test, and can also contribute to the infant mortality

---

[1] Screens are defined as tests that subject 100% of the parts to the test conditions.

failures. The fact that functional and ATPG tests are not perfect also contributes to a high probability of escaped defects. For example, these tests do not detect failure modes that do not result in single stuck-at faults. Such failure modes include bridging to adjacent nets (since they may not held at a zero or one during test) and leakage current.

15.2 Burn-In Tests

Burn-in is a microelectronic part-level screen performed to precipitate defects by exposing the parts to accelerated stress levels, such as temperature and voltage. The goal is to prevent failures from occurring in the field [Pecht et al. 1995].

Burn-in requirements were instituted during the time of the Minuteman Missile Program where the screen was shown to be effective in uncovering defects in low-volume immature parts. By 1968, burn-in was incorporated in military standard MIL-STD-883 [MIL-STD-883 1968], which is in use today.

Burn-in processes commonly consist of placing parts in a thermal chamber for a specific amount of time under an electrical bias. During and/or after thermal environmental exposure, functional tests are conducted. Parts that fail to meet the device manufacturer's specifications are discarded; parts that pass are used.

Burn-in is an accelerated life test that consists of placing products in a thermal chamber for a period of time. An elevated temperature is employed to accelerate the precipitation of infant mortalities. For example, a one-week burn-in test at 125°C is assumed to be equivalent to one year of field life. Functional/ATPG tests performed after (and sometimes during) burn-in are used to screen out the products that do not meet specifications. To deliver adequate stress to all nodes on a chip, products must be subjected to a selected set of test vectors so that a maximum number of nodes are forced into the desired state. Products that pass the test are shipped; those that fail are discarded. Additional burn-in procedures, including exposure to humidity, vibration, temperature and power cycling, are sometimes employed to improve the effectiveness of the screen, raise the acceleration factor, and reduce the burn-in duration.

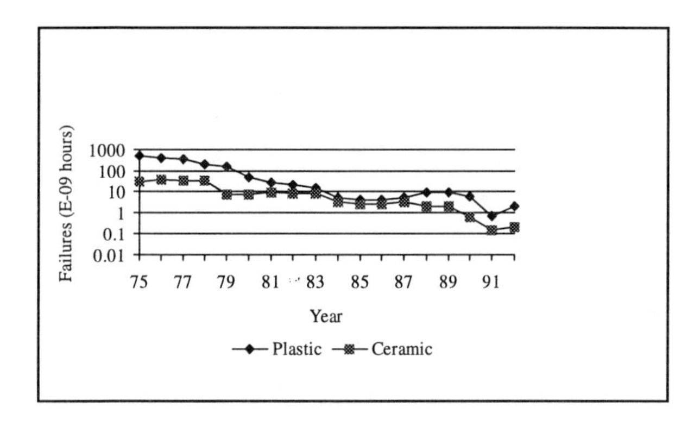

Figure 15-1 125 °C Burn-In Results

The interested reader can find more information on these methods in the book "Quality Conformance and Qualification of Microelectronic Packages and Interconnects" [Pecht et. al. 1994].

When performing burn-in, it is important that the screening duration and stress magnitudes be determined using engineering judgment. Traditionally, these parameters have been based on the bathtub curve failure pattern. However, other failure patterns, such as the modified bathtub curve or the roller-coaster curve, can be used to plan a burn-in procedure. The use of outdated specifications, often based on obsolete failure patterns, will not provide an accurate assessment of current realities. As a result, the useful life of the product may be reduced, good products may be damaged, and some defects may not be screened. Another disadvantage to burn-in is its potential to induce failures due to the inherent additional handling. Such damages include lead bending, package chipping, electrostatic discharge and electrical overstress. The defects introduced during burn-in can be significant relative to the defects that might be precipitated. Therefore, the decision to burn-in, and the duration and stress levels of burn-in, must be thoroughly considered and continuously re-evaluated. Over the last decade there has been scattered evidence that burn-in does not precipitate many defects. For example, Figure 15.1 shows decreasing failure rates precipitated during burn-in testing over the past twenty years [Slay 1995]. In 1990 Motorola wrote that, "The reliability of integrated circuits has improved considerably over the past five years. As a result, burn-in prior to usage does not remove many failures. On the contrary it may cause failures due to additional handling" [Harry, 1990]. In 1994, Mark Gorniak of the U.S. Air Force stated that, "these end-of-line screens [reference Mil-Std-883] provide a standard series of reliability tests for the industry. Although manufacturers continue to use these screens today, most of the screens are impractical or need technology [updates]" [Gorniak 1994].

Table 15-1 Companies Who Don't Implement Burn-In

Company	Product
Hewlett Packard	PC motherboards
APCD, AHMO, IPO	LAN cards, printer cards, SIMMS
Seagate, Singapore Tech.	Disk drive cards
Compaq Asia	Modem cards
TI, NEC Semiconductors	SIMM modules
Exabyte	Tape drive cards
Baxter	Infusion pump PCBA's
Apple	Video tuner cards

Furthermore, in a recent U.S. government funded trip to Singapore and Malaysia taken by six leading experts in the electronics industry, it was found that many companies do not use burn-in on various products (Table 15.1) due to the resulting reduced reliability [Pecht et al. 1996].

This section presents findings on the effectiveness of burn-in for precipitating failures. Specific problems associated with burn-in are discussed, along with the expense associated with performing burn-in. A comparison of the field reliability between unscreened and screened products is then presented. The conclusion provides an alternative approach to the current widespread philosophy of screening and screening elimination.

15.2.1 Burn-In Data Observations

Burn-in data from six companies; National Semiconductor, Motorola, Third Party Screening House, Air Transport Systems Division's (ATSD) Third Party Screening House, Honeywell and Texas Instruments are presented in Tables 15.1-15.8. The data consist of the total number of parts burned-in along with the number of apparent failures detected via functional tests. Apparent failures are classified as either non-valid or valid. Valid failures are those that would have occurred in the field if burn-in had not been performed. Non-valid failures are those that occurred due to handling or other problems that are unique to the burn-in process and thus would not have occurred if burn-in was not performed.

National Semiconductor burn-in data (Table 15.2) show that of the 1,119 parts exposed to burn-in conditions, 42 (3.8%) resulted in apparent failures [Defense Logistics Agency 1995]. The apparent failures were due to 35 mechanical rejects and 7 that retested OK. Burn-in did not precipitate any valid failures.

Table 15-2 National Semiconductor Burn-In Data

Group #	Total Parts Screened	Apparent Failures	Valid Failures
1	347	11	0
Group #	Total Parts Screened	Apparent Failures	Valid Failures
2	452	24	0
3	160	2	0
4	160	5	0
Total	1,119	42	0

Table 15-3 National Semiconductor Burn-In Data

Total Parts Screened	Apparent Failures	Valid Failures
169,508	6	1

Table 15-4 Motorola Burn-In Data

Total Parts Screened	Apparent Failures	Valid Failures
257,908	186	0

Another study conducted by National Semiconductor shows that the burn-in screen (Table 15.3), consisting of 169,508 parts, resulted in 6 (0.0035%) apparent failures. Five (83%) of these failures were due to electrical overstress (EOS) and electrostatic discharge (ESD) damage. One (17%) failure was a valid failure and was due to AC propagation delay.

Motorola burn-in data (Table 15.4) show that of 257,908 parts exposed to burn-in conditions, 186 (0.072%) apparent failures resulted [Defense Logistics Agency 1995]. The apparent failures were due to 182 electrical rejects and 4 mechanical rejects. Of the apparent failures, none were valid.

A third party burned-in 6,105 parts that resulted in 167 (2.7%) apparent failures (Table 15.5) [Defense Logistics Agency 1995]. The apparent failures were due to 143 mechanical rejects and 24 electrical rejects that were caused by testing errors at the screening facility. Of the apparent failures, none were valid.

Table 15-5 Third Party Burn-In Data

Total Parts Screened	Apparent Failures	Valid Failures
6,105	167	0

Table 15-6 ASTD Third Party Burn-In Data

Group #	# Parts Screened	Total Rejects	Valid Failures
1	1,550	12	0
2	2,534	108	1
3	8,548	275	0
4	6,221	237	0
Total	18,853	632	1

Burn-in results from another third party are shown in Table 15.6. ATSD standards were used whenever a burn-in failure was discovered by a post burn-in electrical test [Defense Logistics Agency 1995]. This test was performed at the same temperature as the pre-burn-in electrical test. In this case, post electrical tests were performed at the upper extreme. Therefore, only post electrical hot failures were failure analyzed and may be considered valid failures. This company burned-in 18,853 parts of which 632 (3.4%) were rejected. The rejects were due to mechanical defects (569), non-burn-in electrical results (13) and burn-in electrical failures (7). The test lab retained 35 parts and 8 parts were lost. Of these total rejects, only one was a valid failure whose root cause was traced to a defect in the fabrication process.

Honeywell burned-in a total of 162,940 parts, of which 669 resulted in apparent failures (see Table 15.7) [Scalise 1996]. Out of 67 parts that were failure analyzed, five were valid failures; two that were process related and three were temperature related. The remaining 62 were invalid failures, where electrical overstress (EOS) and ESD contributed to 49 (73%) of the failures.

Table 15-7 Honeywell Burn-In Data

# Parts Screened	Apparent Failures	Valid Failures
162,940	67	5

Table 15-8 Texas Instrument Burn-In Data

Technology	# Parts Screened	Total Failures
TTL	54,848	5
S	63,347	13
LS	76,865	7
Total	195,070	25

Texas Instruments (TI) burned-in a total of 195,070 different TTL, S and LS parts (Table 15.8) [Tang 1996]. Of these parts, 25 displayed apparent failures. Two valid failures resulted, one due to a die mechanical damage and the other from a broken wirebond. Another failure could not be resolved. The 22 remaining apparent failures were due to EOS and ESD damage.

Texas Instruments data of HCMOS (Table 15.9) show that out of the 100,165 parts that were burned-in, 8 resulted in apparent failures [Scalise 1996]. All of these failures were due to EOS or ESD.

Table 15-9 Texas Instruments HCMOS Burn-In Data

Technology	# Parts Screened	Total Failures
HCMOS	100,164	8

15.2.2 Discussion of Burn-In Data

Of the total 911,667 parts, the data presented show that burn-in detected 1125 (0.12%) apparent failures of which 1116 were invalid and 9 were valid. That is, valid failures were found on only 0.001% of the parts and only 0.8% of the apparent failures. The valid failures consisted of: AC propagation delay, a defect in the fabrication process, die mechanical damage, and a broken wirebond.

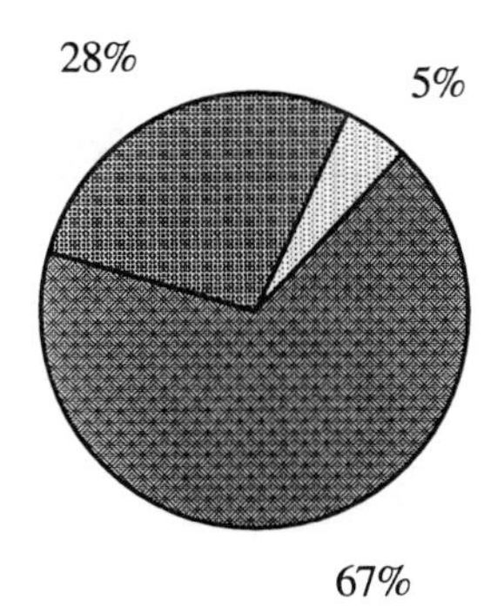

Figure 15-2 Non-valid failures

The breakdown of non-valid failures is shown in Figure 15.2. Mechanical defects included improper device orientation and bent leads. Electrical rejects include ESD and EOS. Additional handling causes ESD, whereas EOS occurs due to misapplied power. The “other” category includes parts that retest OK and parts that are retained by the test lab or lost. Retest OK is defined as parts that do not pass the functional test initially, but do pass in a subsequent test. These failures are not device related. For example, a contact may have dust contamination and when taken out of its socket and re-inserted, the dust particles are removed creating a better contact that enables the device to pass the functional test. Data obtained from Northbrook, in 1991-1992, showed that of 1,017,828 parts burned-in, 70% of the detected defects were due to wafer processing and the remaining defects were package related [Wager et al. 1973]. In 1993-1994, 582,480 parts (from the same manufacturers) were burned-in, with 100% of the detected defects due to the wafer process. This suggests that package quality has improved to the point where burn-in is essentially non-value added. In terms of this study, because of the extremely small percentage of valid failures, it is difficult to compare the effectiveness of burn-in to precipitate die level versus package level defects. The key point in our collected data is that, by the definition of invalid, 99.2% of the failures could have been avoided if burn-in was not performed. Further, as will be discussed in Section 15.2.4, additional damage may have occurred due to the additional handling inherent in burn-in screening; these failures can be precipitated in the field application.

Table 15-10 Burn-In Expenses

Part Cost ($)	Burn-In Cost ($)	Total Cost ($)	% Burn-In Costs
45.00	23.00	68.00	33.82%
11.15	18.35	29.50	62.20%
2.50	6.25	8.75	71.43%
1.02	6.48	7.50	86.40%
0.90	0.80	1.70	47.06%
0.80	5.45	6.25	87.20%
0.80	0.41	1.21	33.88%
0.50	1.85	2.35	78.72%
0.28	13.52	13.80	97.97%
0.21	0.18	0.39	46.15%
0.11	1.60	1.71	93.57%

15.2.3 Burn-In Costs

Table 15.10 presents actual burn-in cost values from a typical screening house. Screening houses charge on a piece basis, a lot basis, or a combination of the two. The costs vary depending on the quantity and complexity of the parts. A complex part may require that more functional tests be performed, whereas a simple part would require basic DC parametric functional testing. The costs also depend on volume. The cost of burn-in is independent of the part cost since the majority of the test costs depend on the complexity of the test program that must be developed.

15.2.4 Higher Field Reliability Without Screening: A Case Study

Honeywell's Air Transport Division has historically screened plastic encapsulated microcircuits (PEMs) for 160 hours at 125°C, followed by a tri-temp screen at -40°C, room temperature and 125°C, in accordance with the traditional view that this test and these temperatures would increase the reliability of the end product due to reduced infant mortality. This practice has always been a questionable activity because:

- Typically the integrated circuit (IC) manufacturer does not perform this screen and thus this third party action becomes suspect and expensive.

- The improvements that semiconductor manufacturers have made in product quality must be addressed with respect to the effect of burn-in. That is, the ability of screening to actually decrease field reliability due to part damage occurring during the screening process must be considered.

In the previous sections, it has been shown that many parts failed during burn-in; the majority of these failures were invalid. Two questions to be posed are: were the parts that failed really defective, and were the parts that were sent to the field reliable.

The first question has been discussed in section 13.2.2, where it was shown that burn-in caused over 99% of the apparent failures; that is, less than 1% of the parts were really defective. This section attempts to answer the second question.

From previous results, it was believed that the handling required to burn-in parts was causing unacceptable ESD damage. What is particularly unsettling with this conclusion is that it raises the issue of latent ESD damage in fielded equipment. The handling required to burn-in parts seems to be causing unacceptable ESD damage. In order to reach a data driven conclusion as to the necessity of part screening, particularly burn-in, data was collected from two sets of data:

- Aircraft field failure data for both military ceramic and commercial plastic parts; where most of the plastic parts were screened, but with a significant and identifiable group that were not screened.

- A ring laser gyro that was built with totally unscreened commercial parts to allow an on-aircraft evaluation of the effects of not screening.

From the first set of data, an examination of the field failures that occurred with the burned-in PEMs are shown in Table 15.11. The results show that of the total failures, 6.6% were valid, 31.7% were invalid, 4% could not be determined and the remaining 57.7% were not failure analyzed. The failures that resulted could be due to the part, sub-system, or system level. For example, the invalid failures do not pertain to those caused by burn-in, as discussed earlier. These invalid failures may be due to a lead that was not soldered to the printed circuit board. A process related failure, which is considered to be a valid failure, could be a solder joint that did not wet properly during the soldering stage.

Of the invalid failures, 6.2% were due to ESD damage. This is far higher than expected based on historical in-service failure data. From this, the conclusion is that burn-in not only adds no value, but it may even increase the field failure rate of the devices built with these burned-in parts. It was also hypothesized that latent ESD effects could be introduced somewhere in the build process of the equipment.

Table 15-11 Field Failure Results: Ring Laser Gyro With Screening

Type Failure Type	Digital SSM/ MSI	Digital LSI/ Mem	Linear	Total Parts Failed	Percent of Total Failures
Valid					
Fabrication	0	2	2	4	1.8
Temperature	0	0	11	11	4.8
Invalid					
EOS/ESD	8	1	5	14	6.2
Other	7	18	33	58	25.5
Undetermined	2	1	6	9	4.0
No failure analysis	16	70	45	131	57.7
Total Parts Failed	33	92	102	227	

To assess this hypothesis, a decision was made to include in the Honeywell product mix, a ring laser gyro that was built with totally unscreened, commercial parts (second data set). The part types and manufacturers used in this assembly were the same as those used in the devices that had the higher ESD related failure rate, i.e., the first data set. This is because the design engineers are required to work from a relatively small list of approved parts and manufacturers. The build facility was also common to all the devices. What this means is that a comparison can be made between devices built with screened parts and devices built with unscreened parts where the only difference is the part screening. These devices are installed in the same aircraft and in the same equipment bays.

The devices built using the unscreened commercial parts have accrued well over 200 million piece part hours *without a failure*. The devices that were screened resulted in 669 (0.4%) failures out of the total 162,940 parts. The only difference between these devices is that screening was not done on the devices that had no failures.

What has been shown in this comparison study is that a high percentage of latent ESD failures resulted when the parts were screened and no failures resulted when the parts were unscreened. Therefore, the parts sent to the field were not as good as originally thought, since burn-in degraded part reliability. That is, burn-in precipitated many invalid failures and degraded part reliability, resulting in field failures. For these reasons, many field failures can be avoided if burn-in is not performed.

15.2.5 Burn-in Recommendations

Many companies involved with "critical systems" require burn-in because they believe that the risk to do otherwise is too high. This so-called safety net viewpoint has become a trap that should be avoided. Burning-in parts from a quality manufacturer will increase the number of field failures. The failures due to the burn-in process can be significant relative to those due to inherent defects and, therefore, an alternate approach to burn-in is necessary.

Since burn-in has often been shown to be ineffective, many companies have begun implementing a burn-in elimination program. This program is often based on accumulated burn-in data and subsequent low failure rates. This practice implies that burn-in is being performed at some point in the process. Our recommendation is that, instead of conducting burn-in and then implementing a burn-in elimination program, manufacturer part family assessment and qualification data should be used to assess the need for burn-in. This approach is based on existing qualification and reliability monitoring data, rather than on burn-in data, thereby avoiding associated part degradation

Manufacturer part family assessment is dependent on the supplier. Parts must come from a supplier that is periodically certified, implements statistical process control (SPC) or an acceptable documented process, has acceptable qualification testing results, abides by procedures to prevent damage or deterioration (e.g., handling procedures, such as ESD bags), and provides change notifications. Once a quality part is obtained, results from the qualification tests can be used to determine whether or not burn-in is warranted. If no failures occur in qualification tests and the manufacturing processes are in control, then confidence can be gained to assess the part quality without performing burn-in.

15.3 IDDQ Tests

IDDQ tests are not based on verifying functionality or detecting hard stuck-at faults, but are instead focused on detecting a type of defect, regardless of how it manifests itself within a given circuit.

The IDDQ test is powerful and does not require a large number of vectors to detect potential defects. Test patterns are required to precondition the component and to put all the circuit nodes in either state. In addition, there is little IDDQ dependence on device design. All that is required is that there are no static pull-ups or pull-downs and that contention states can be turned off.

IDDQ testing is capable of detecting inter-level shorts, multiple stuck-at-faults, pinholes, spot defects and soft defects [Al-Assadi, 1996]. IDDQ can identify a wide range of failure types including those undetectable by stuck-at-fault testing. Such types include parasitic transistor leakage, gate-oxide defects, bridging defects, leaky p-n junction, and transistor faults [Al-Assadi, 1996; Wantuk, 1996]. These faults have a tendency to cause elevated quiescent power supply current, which is detectable by IDDQ. The effectiveness of the IDDQ screen is measured through IDDQ coverage, defined here as the percentage of nodes held in both states (logical "0" and "1") during IDDQ measurements.

The disadvantage of IDDQ-based tests is that it is extremely difficult to isolate the location of a fault and to identify the root cause. The definition of the specification limit (e.g., how much current is too much) is subjective. Finally, IDDQ measurements tend to be slow, so that a large number of IDDQ vectors increases the test time.

Consequently, IDDQ tests are typically used in conjunction with stuck-at or functional tests [Baker 1996 Wantuk 1996]. Availability of both types of tests tends to produce a high quality test while optimizing the test time. The combination of both tests has the capability of detecting potential, as well as hard defects. However, fault isolation and some form of functional testing will still be necessary.

16. SUMMARY

There are a number of different mechanisms that impact IC component reliability, as summarized in Table 16.1. The failures are caused by material wear-out due to the high stresses associated with their use in ICs, but they are also manifested in all three portions of the bathtub curve, mostly due to interactions with defects and/or design sensitivities.

Most of this book focuses on the issues associated with the individual failure mechanisms, and hence may convey a sense of high risk. However, it is clear that the industry practices evolved to date manage all these reliability issues very well. Modern systems, containing many millions of transistors, are, by and large, very reliable, and tend to stay up longer and perform far better than was thought possible only a few years ago. The issue then is what happens in the future – how is the understanding of the failure mechanisms gained to date, as summarized in this chapter, going to be leveraged in the future? It is clear that laws of physics will apply to future products, too, and in fact that they will be 'more strictly enforced'. It is also clear that some of the practices from the past will have to be changed to deal with the future

Table 16-1 Reliability Mechanisms, Failure Modes and their Key Reliability Indices

Failure Mechanism	Failure Mode	Reliability Index
Electromigration	opens	area of minimum width metal, number of inter-level metal overlaps
	shorts	area of inter-level metal overlaps number of inter-level metal overlaps
		area of minimum spaced metal
	contact spikes	number of diffusion contacts
	via spikes	number of vias
Latch-up	shorts on the die	area of minimum spaced wells
Leakage current[2]	parametric shift in device	collector-base current with emitter open

---

[2] Leakage current is not a failure mechanism by itself, although it can lead to uncontrolled increase in junction temperatures.

Failure Mechanism	Failure Mode	Reliability Index
Hot Carrier Aging	electron entrapment	substrate current degradation, photoemission intensity
Self-Heating[3]	snapback (thermal or second breakdown) in BJT	voltage drop
	distortion of small signal parameters in BJT	frequency variation
	non-linear transconductance in large signal BJT	increase in propagation delay time
Transconductance Degradation	oxide trapping	drain voltage, gate geometry, carrier mobility
	slow trapping	aluminum sintering
Transconductance Degradation	substrate defects	radiation flux
Electrostatic Discharge	bridging	area of minimum width metal, number of inter-level overlaps
	opens	area of minimum width metal, number of inter-level overlaps
	gate oxide breakdown	oxide thickness, area of NMOS gate oxide, area of PMOS gate oxide, length of gate-diffusion edges, number of gate-diffusion overlaps

[3] Self-Heating is not a failure mechanism by itself. It is a mode of heat generation and concentration at the junction level which can and does precipitate other failures at the device level which are temperature dependent.

Clearly, deep sub-micron technologies will aggravate many of the reliability mechanisms and the way that they are managed. However, the physical foundation behind most of these trends is based on increasing chip complexity, increasing chip operating frequency, and decreasing power supply voltage.

These factors are the source of the technical concerns posed by the deep sub micron technologies, as summarized in Table 16.2.

Managing typical specifications that define chip performance, cost, schedule, quality and reliability becomes significantly more difficult with deep sub micron chip complexities. It is clear that the drive throughout the industry is toward developing design tools and practices that manage this balance better.

Table 16-2 Deep Sub-Micron Challenges

Timing	• foil delay becoming dominant over MOS delays • physical layout (routing) a big factor • more difficult to manage due to complexity
Cross Talk	• tight multi-level spaces in long metal runs
Tight Si Window	• process sigmas are a larger % of mean
Signal Integrity	• reduced signal to noise ratios
Switching Noise	• increased number of pins & signals
Power Dissipation	• increasing due to chip complexity & frequency
Testability	• difficult due to chip complexity • need to keep test time and tester cost down • problem with mixed signal/analog products
Debugability	• very complex multi-million gate designs • difficult with imported intellectual property • hierarchical designs • limited observability due to BIST trends
Reliability	• difficult due to de-layering trends • wear out closer due to sheer stress levels • infant mortality higher due to sensitivity • failure rate higher due to complexity & size

Thus, for example, true 'timing driven' place and route tools, capable of handling all the performance issues automatically off some general specification, are the holy grail of the EDA business. Such focus on design for reliability does not exist, because lack of it is not perceived as a show stopper. But, it is growing. It is clear that more design for reliability is needed, and that more design for reliability is being implemented. And, application of the knowledge of the fundamental physics of failures, as summarized here, as well as an understanding of the design and business realities associated with the IC industry, are the paths to more and better design for reliability practices. In addition, it is clear that the philosophy of design for reliability must also encompass concurrent engineering practices, where all sources of information are leveraged at the earliest phases in product development.

REFERENCES

Adams, A. C. et al., "Characterization of plasma-deposited silicon dioxide," *J. Electrochemical Society: Solid State Science and Technology*, Vol. 128, No. 7, 1981.

Ahlburn, B. et al., "Advanced dielectric techniques for the fabrication of 16 megabit DRAM generation devices," *Proc. 3rd Intl. Symp. On ULSI Sci. and Tech.*, ECS, Vol. 91-11, 1991.

Al-Assadi, W.K., Jayasumana, A.P., and Malaiya, Y.K. "A Bipartite, Differential IDDQ Testable Static RAM Design," *Technical Report CS-96-101*, Colorado State University, Dept. of Electrical Engineering/Dept. of Computer Science, February, 1996.

Al-Assadi, W.K., Jayasumana, A.P., and Malaiya, Y.K., "Differential IDDQ Testable Static RAM Architecture," *Technical Report CS-96-102,* Colorado State University, Dept. of Electrical Engineering/Dept. of Computer Science February, 1996.

Amerasekera, A. and Campbell, D.S., *Failure Mechanisms in Semiconductor Devices*, Wiley, New York, 1987.

Amerasekera, A., Chang, M., Seitchik, J., Chatterjee, A., Mayaram, K., and Chern, J., "Self-Heating Effects in Basic Semiconductor Structures," *IEEE Transactions on Electron Devices,* vol. 40, pp. 1836-1844, 1993.

Arita, Y. et al., "Deep submicron Cu planar interconnection technology using CU selective chemical vapor deposition," *IEDM*, 1989.

ASM International Handbook Committee, *Electronic Materials Handbook: Packaging*, ASM International, Materials Park, OH, 1989.

Baker, K., Marinissen, E.J., and Bruls, E., "QTAG Industry Standards for IDDQ Testing," *Philips IDDQ Test Process for CDF-AEC*, AEC Conference, 1996.

Bakker, D., "Applied use of advanced inspection systems to measure, reduce, and control defect densities," *SPIE*, Vol. 1261, *Integrated Circuit Metrology, Inspection, and Process Control IV,* 1990.

Bar-Cohen, A., "Reliability Physics vs Reliability Prediction," *IEEE Transactions on Reliability,* vol 37, p 452, December 1988.

Bartelink, D. J. and Chiu, K. Y., "Interconnects for submicron ASIC's," *Intl. Symp. On VLSI Tech. Sys. and Appl.*, 1989.

Bateson, J., 1985, *In-Circuit Testing*, Van Nostrand Reinhold, New York.

Becker, F. S. and Rohl, S., "Low Pressure deposition of doped SiO2 by pyrolysis of tetraethylorthosilicate (TEOS)," *J. Electrochem. Soc. Solid State Science and Technology*, November 1987.

Becker, F. S. et al., “Process and film characterization of low pressure tetraethylorthosilicate borophosphosilicate glass,” *J. Vac. Sci. Technol.*, B4 (3), May/June 1986.

Billington, D. and Crawford, J., *Radiation Damage in Solids*, Princeton Univ. Press, NJ, 1961.

Blower, R. S., “Progress in LPCVD tungsten for advanced microelectronics applications,” *Solid State Technology*, 1986.

Blum, J. M., “Chemical vapor deposition of dielectrics: A review,” *Proc. of the 10th Intl. Conf. On CVD*, 1987.

Bowers, G.H., “Continuous flow manufacturing,” *Intl. Semiconductor Manufacturing Symp.,* 1990.

Broadbent, E. L. and Stacey, W. T., “Selective tungsten processing by low pressure CVD,” *Solid State Technology*, December 1985.

Bromstead J. R. and Bauman E. D., “Performance of Power Semiconductor Devices at High Temperature,” *First International High Temperature Electronics Conference,* Alberqerque, NM, June 1991.

Brown, D. M. et al., “Trends in advanced process technology-submicrometer CMOS device design and process requirements,” *Proc. IEEE*, Vol. 74, No. 12, December 1986.

Buckroyd, A, *In-Circuit Testing*, Butterworth-Heinemann Ltd., Boston, MA, 1994.

Burba, M. E. et al., “Selective dry etching of tungsten for VLSI metallization,” *J. Electrochem Soc.*, October 1986.

Burggraaf, Peter, “Thin film technology for advanced semiconductors, part 2b: Dielectrics,” *Semiconductor International*, July 1985.

Burmester, R. et al., “Reduction of titanium silicide degradation during borophosphosilicate glass reflow,” *ESSDERC*, 1989.

Cagnoni, P. et al., “A triple metal interconnection process for CMOS technology,” *VMIC*, June 1989.

CALCE Electronic Packaging Research Center, “The Use of Electronic Hardware at Elevated Temperatures to 200°C,” Internal report, University of Maryland, 1996.

Caludius, H., “Practical defect reduction in an MOS IC line,” *Microcontamination,* April 1991.

Cameron, G. and Chambers, A., “Successfully addressing post-etch corrosion,” *Semiconductor International*, May 1989.

Carpio, R. A., “IR spectroscopy enhances dielectric film QC,” *Semiconductor International*, August 1989.

Castel, E. and Ray, A., "An integrated approach to defect detection, analysis and reduction in photolithography," *Microelectric Engineering,* Vol. 6, 1987.

Cemal, D. T. et al., "Hot Carrier Degradation in LDD-MOSFETs at High Temperatures," *Solid-State Electronics*, Vol. 37, No. 12, pp. 1993-1995, 1994.

Chan, T.Y. and Ko, P.K., "A Simple Method to Characterize Substrate Current in MOSFETs," *IEEE Electron Device Letters,* Vol. 5, No. 12, pp. 505-507, 1984.

Cheek, R. W. et al., "Methanol passivation of SiO2 for selective CVD tungsten," *Advanced Metallization Applications*, University of California, Berkeley, 1991.

Coburn, J. W. and Winters, H. F., "Plasma Etching: a discussion of mechanisms" *J. Vac. Sci. Technol.*, 16, 1979.

Coburn, J. W., "Plasma etching and reactive ion etching," *American Vacuum Society*, 1982.

Coleman, R. and Chitturi, P., "Defect reduction strategies for submicron manufacturing tools and methodologies," *SPIE*, Vol. *1392, Advanced Techniques for Integrated Circuit Processing*, 1990.

Comello, V., "Planarization using RIE and chemical mechanical polish," *Semiconductor International*, March 1990.

Cooper, S. and Geisinger, C., "Inspection improves production yield," *Electronic Packaging and Production,* March 1988.

Cunningham, J., "Using the learning curve as a management tool," *IEEE Spectrum,* June 1990.

Cushing, M. J., Mortin, D. E., Stadterman, T. J., and A. Malhotra, "Comparison of Electronics Reliability Assessment Approaches," *IEEE Transactions on Reliability,* Vol. 42, pp. 600-607, December, 1993.

Cushing, M. J., Stadterman, T. J., Krowelski, J. T., and Hum, B. T., "U.S. Army Reliability Standardization Improvement Policy and Its Impact," *IEEE Transactions on Components, Packaging, and Manufacturing Technology,* Part A, Vol. 19, No. 2, June 1986.

Dally, J., *Packaging of Electronic Systems*, McGraw-Hill, New York., 1990

Dance, D. and Jarvis, R., "Using yield models to accelerate the learning curve progress," *IEEE Trans. on Semiconductor Manufacturing*, February 1992.

Daubenspeck, T. et al., "Planarization of ULSI topography over variable pattern densities," *J. Electrochem. Soc.*, Vol. 138, 1991.

Davari, B. et al., "A new planarization technique using a combination of RIE and chemical mechanical polish (CMP)," *IEDM* 89-61, 1989.

Davis, D. B., "Technologies ride a fast track into the 1990s," *Electronic Business,* December 1989.

De Santi, G., "Interconnections technologies for VLSI circuits," *ETT*, Vol. 1, No. 2, March-April 1990.

Defense Logistics Agency, Plastic Package Availability Program, Final Report, Contract Number DLA900-92-C-1647, Sharp Proceedings, Indianapolis, IN, November 15, 1995.

Department of Defense, *MIL-STD-883:Test Methods and Procedures for Microelectronics,* 1968.

Dimitrijev, S. et al., "Yield model for in-line integrated circuit production control," *Solid State Electronics,* Vol. 31, No. 5, 1988.

Donovan, R. P. (ed.), *Particle Control for Semiconductor Manufacturing*, Marcel Dekker, New York, 1990.

Electronic Materials Handbook, Packaging, ASM International, Materials Park, vol. 1, OH, 1989.

Elliot, J. K., "Current trends in VLSI materials," *Semiconductor International*, March 1988.

Emesh, T., "PECVD of silicon dioxide using tetraethylorthosilicate (TEOS)," *J. Electrochem. Soc.*, Vol. 136, No. 11, November 1989.

Ephrath, L. M. and Mathad, G. S., "Etching-application and trends of dry etching," in G. Rabbat (ed.), *Handbook of Advanced Semiconductor Technology and Computer Systems*, Van Nostrand Reinhold Co., 1988.

Feldman, P. and Director, S., "A macromodeling based approach for efficient IC yield optimization," *Research Report No. CMUCAD–91-27,* Carnegie-Mellon University, 1991.

Feldman, P. and Director, S., "Accurate and efficient evaluation of circuit yield and yield gradients," *Research report No. CMUCAD-90-32,* Carnegie-Mellon University, 1990.

Feldman, P. and Director, S., "Improved methods for IC yield and quality optimization using surface integrals," *Research Report No. CMUCAD-91-31,* Carnegie-Mellon University, 1991.

Fisher, D. et al., "A submicron CMOS triple level metal technology for ASIC applications," *IEEE Custom Integrated Circuits Conf.,* 1989.

Flack, V., "Estimating variation in IC yield estimates," *IEEE J. Solid State Circuits,* Vol. SC-21, No. 2, 1986.

Fox, R., Lee, S., and Zweidinger, D., "The Effects of BJT Self-Heating on Circuit Behavior," *IEEE Journal of Solid State Circuits,* Vol. 28, pp. 678-685, 1993.

Friedman, D. and Albin, S., "Clustered defects in IC fabrication: Impact on process control charts," *IEEE Trans. On Semiconductor Manufacturing*, February 1991.

Fritzsche, H. et al., "An improved etchback process for multilevel metallization and its reliability results for CMOS devices," *VMIC*, June 1986.

Geraghty, P. and Harshbarger, W. R., "Metal interconnects for ULSI integrated circuits," *Proc. Tech. Prog. Nat. Elect. Packg. Conf.*, NEPCON West, 1990.

Ghandhi, S. K., VLSI Fabrication Principles, John Wiley & Sons, 1983.

Glang, R., "Defect size distributions in VLSI chips," *IEEE Trans. On Semiconductor Manufacturing,* November 1991.

Gorniak, M., Rome Labs, Qualified Manufacturers List update, *Reliability Society Newsletter*, Vol. 40, p. 8, 1994.

Gorowitz, B. et al., "Recent trends in LOCVD and PECVD," *Solid State Technology*, October 1987.

Gupta, S. K., "Spin-on-glass films in multilevel IC interconnection," *Mat. Res. Soc. Symp. Proc.*, Vol. 108, 1988.

Gupta, S. K., "Spin-on-glass for dielectric planarization," *Microelectronic Manufacturing and Testing*, Vol. 12, No. 5, April 1989.

Gupta, S. K., et al., "Materials for contacts, barriers, and interconnects," *Semiconductor International*, March 1990.

Guthrie, W. L. et al., "Chemical mechanical polishing to planarize blanket and selective CVD tungsten," *Advanced Metallization Applications*, University of California, Berkeley, October 1991.

Harry, M. J., "The Nature of Six Sigma Quality," *Motorola Handbook*, pp. 1-25, 1990.

Henderson, I., "A production fab defect reduction program," *Intl. Semiconductor Manufacturing Sci. Symp.,* 1989.

Hey, H. P. W. and Machado, J. R. S., "An oxide step coverage comparison for different plasma CVD reactions (disilane/N2, silane /N2O, and TEOS/O2)," *Proc. 175th Electrochem. Soc. Meeting*, Vol. 89, No.1, May 1989.

Hey, H. P. W. et al., "Ion bombardment: A determining factor in plasma CVD," *Solid State Technology*, April 1990.

Hey, H. P. W., "Progress in plasma CVD technology," *Microelectronic Manufacturing and Testing,* January/February 1990.

Hills, G. W, et al., "Plasma assisted deposition and device technology: Interlevel dielectric consideration," *SPIE: Dry Processing for Submicrometer Lithography*, Vol. 1185, 1989.

Hnatek, E. R., *Integrated Circuit Quality and Reliability*, Marcel Dekker, Inc., New York, 1995.

Hoyt, B., *The Implementation of Statistical Process Control (SPC),* Northcon, 1987.

Hsu, F.C. and Chiu, K.Y., "Temperature Dependency of Hot-Electron-Induced Degradation in MOSFETs," *IEEE Electron Device Letters.* Vol. 5, pp. 148-150, 1984.

Hu, C. et al., "Hot-Electron-Induced MOSFET Degradation - Model, Monitor, and Improvement," *IEEE Transactions. of Electronic Devices,* Vol. 32, No. 2, pp. 375-385, 1985.

Hu, C. K. et al., "Dry etching of TiN/Al(Cu)/Si for very large scale integrated local interconnections," *J. Vac. Sci. Technol.*, A 8 (3), May/June 1990.

Hurley, K. H. et al., "BPSG films deposited by APCVD," *Semiconductor International*, October 1987.

Hurley, K. H., "Process conditions affecting boron and phosphorous in borophosphosilicate glass as measured by FTIR spectroscopy," *Solid State Technology*, March 1987.

Ibbotson, D. E. et al., "Oxide deposition by PECVD," *SPIE, Monitoring and Control of Plasma-Enhanced Processing of Semiconductors,*" Vol. 1037, 1988.

Ito, S. et al., "Application of surface reformed thick spin-on-glass to MOS device planarization," *J. Electrochem. Soc.*, Vol. 137, No. 4, April 1990.

Jaso, M. A. et al., "Etch selectivity of silicon dioxide over titanium silicide using CF4/H2 reactive ion etching," *J. Electrochem. Soc.*, Vol. 136, No. 7, July 1989.

Johnson, P. B. et al., "Using BPSG as an interlayer dielectric," *Semiconductor International*, October 1987.

Jones, W.K. and Landis, D., "Instrumentation and Testing," *Electronic Materials Handbook,* Vol. 1, pp. 377-378, 1989.

Jordan, J. and Pecht, M., "How Burn-In Reduces Quality and Reliability," *Journal of the Institute of Environmental Sciences*, Vol. 39, 1996.

Jucha, B. and Davis, C., "Reactive ion etching of thick CVD tungsten films," *VMIC*, 1988.

Kamins, T.I. "Structure and stability of low pressure chemically vapor-deposited silicon films," *J. Electrochem. Soc. Solid State Science and Technology*, June 1978.

Kawai, M. et al., "Interlayered dielectric planarization with TEOS-CVD and SOG," *VMIC*, June 1988.

Kern, W. and Kurylo, W. A., "Optimized chemical vapor deposition of borophosphosilicate glass films," *RCA Review*, Vol. 46, June 1985.

Kern, W. and Schnable, G. L., "Chemically vapor deposited borophosphosilicate films for VLSI applications," *Solid State Technology*, January 1984.

Kern, W. and Smeltzer, R. K., "Borophosphosilicate glasses for integrated circuits," *Solid State Technology*, June 1985.

Kim, S. U. and Puttlitz, Albert F., "Plasma process induced device degradation," *IEEE Transactions on Components, Hybrids and Manufacturing Technology*, Vol. CHMT-8, Mo. 4, December 1985.

King, C. et al., "Electrical defect monitoring for process control," *SPIE*, Vol. 1087, IC *Metrology, Inspection and Process Control III*, 1989.

Kojima, H. et al., "Planarization process using a multi-coating of spin-on-glass," *VMIC*, June 1988.

Kotani, H. et al., "Low temperature APCVD oxide using TEOS-ozone chemistry for multilevel interconnections," *IEEE IEDM*, 1989.

Kotani, H., "Metallization for ULSI," *20th Conf. On Solid State Devices and Materials*, 1988.

Kotani, H., "Metallization for ULSI," *Extended Abs. 20th Conf. Solid State Dev. Matl*, 1988.

Kup, Y., "Planarization of multilevel metallization processes: A critical review," *SPIE Advanced Processing of Semiconductor Devices*, Vol. 797, 1987.

Lacy, E. A., "Protecting Electronic Equipment from Electrostatic Discharge," *Tab Professional Reference*, Blue Ridge Summit, PA, 1994.

Lea, R. and Bolouri, H., "Fault tolerance: Step towards WSI," *IEEE Proceedings* Vol. 135, Pt. E, No. 6, 1988.

Lee, P. et al., "Chemical vapor deposition of tungsten (CVD W) as submicron interconnection and via stud," *J. Electrochem. Soc.*, Vol. 136, No. 7, July 1989.

Lee, T.W. and Pabbisetty, S., editors, *Microelectronic Failure Analysis*, ASM International, Materials Park, OH., pp. 301-302, 3rd edition, 1993.

Lifshitz, N. and Smolinsky, G., "Detection of water-related charge in electronic dielectrics," *Appl. Phys. Lett.*, Vol. 55, No. 4, July 1989.

Lifshitz, N. et al., "Water related degradation of contacts in the multilevel MOS IC with spin-on-glass as interlevel dielectrics," *IEEE Electron Device Letters*, Vol. 10, No. 12, December 1989.

Liljegren, D., "Defect reduction strategies for process control and yield improvement," *SPIE,* Vol. 1392, *Advanced Techniques for Integrated Circuit Processing,"* 1990.

Lutze, J. W. et al., "Anisotropic reactive etching of aluminum using CL2, BCl3, and CH4 gases," *J. Electrochem. Soc.*, January 1990.

Magnella, C. G. and Ingwersen, T., "A comparison of planarization properties of TEOS and SiH4 PECVD Oxides," *VMIC*, 1988.

Maluf, N. I. et al., "Selective tungsten filling of sub-0.25 μm trenches for the fabrication of scaled contacts and x-ray masks," *J. Vac. Sci. Technol.*, B 8 (3), May/June 1990.

Maly , W. and Strojwas, A., *SRC Technical Report T91118,* August 1991.

Martinez-Duart, J.M and Albella, J.M, "Metallization technologies for ULSI," *Vacuum*, Vol. 39, No. 7/8, 1989.

Materials Research Society Symposium Proceedings, *Electronic Packaging Materials Science V*, MRS, Pittsburgh, PA, Vol. 203, 1990.

Matsumoto, H. et al., "Effect of long term stress on IGFET degradations due to hot electron trapping," *IEEE Transactions of Electronic Devices*, vol. 28, p. 923, 1981.

May, P. and Spiers, A. I., "Dry etching of aluminum/TiW layers for multilevel metallizations in VLSI," *J. Electrochem. Soc.*, June 1988.

May, T.C. and Murray, H.W., "A new physical mechanism for soft errors in dynamic memories" *IEEE Reliability Physics Symposium*, San Diego, CA, p. 33, 1978.

McCarthy, A. et al., "A novel technique for detecting lithographic defects," *IEEE Transactions on Semiconductor Manufacturing,* Vol. 1, No. 1, February 1988.

McClean, W.J. (ed.) *Status* 1992:*A Report on the Integrated Circuits Industry,* Integrated Circuits Engineering Corp., 1992.

Mercier, J. S., "Rapid flow of doped glasses for VLSIC fabrication," *Solid State Technology*, July 1987.

Michalka, T. et al., "A discussion of yield modeling with defect clustering, circuit repair, and circuit redundancy," *IEEE Trans. on Semiconductor Manufacturing,* August 1990.

Mikel, J.H., The Nature of Six Sigma Quality, *Motorola Handbook*, pp. 1-25, 1990.

Mitsuhashi, K. et al., "Interconnection technology for three-dimensional integration," *Japanese J. App. Phy.*, Vol. 28, No. 4, April 1989.

Moghadam, F., "Planarization developments," *VLSI Multilevel Interconnection State of the Art Seminar*, 1992.

Molnar, L. D., "SOG planarization proves better than photoresist etch back," *Semiconductor International*, August 1989.

Moore, D. and Walker, H., *Yield Simulation for Integrated Circuits,* Kluwer Academic Publishers, 1987.

Moore, W. et al., "Yield modeling and defect tolerance in VLSI," *Intl. Workshop on Designing for Yield,* July 1987.

Nagy, A. and Helbert, J., "Planarized inorganic interlevel dielectric for multilevel metallization-Part-1," *Solid State Technology*, January 1991.

Nguyen, S. et al., "Reaction mechanisms of plasma and thermal assisted CVD of TEOS oxide films," *J. Electrochem.* Soc., 137, 1990.

Nicolet, M. A. "Diffusion barriers in this films," *Thin Solid Films*, 52, 1978.

Nishida, T. et al., "Multilevel interconnection for half-micron ULSI's," *VMIC*, June 1989.

Normandim, M. S. et al., "Characterization of silicon oxide films deposited using tetraethylorthosilicate," *Canadian J. Phys.*, Vol. 67, 1989.

Oakley, R. E., "Advanced interconnect for VLSI," *ESPRIT '84: Status Report of Ongoing Work*, 1985.

Obha, T. et al., "Dimethlyhydrazine surface cleaning and selective CVD tungsten," *Advanced Metallization Applications*, University of California, Berkeley, 1991.

Ohba, T., "Multilevel metallization trends in Japan," *Advanced Metallization for ULSI Applications*, University of California, Berkeley, 1991.

Oshiro, L. and Radojcic, R., A Design Reliability Methodology for CMOS VLSI Circuits," *IEEE International Reliability Workshop Final Report*, Lake Tahoe, CA, 1992.

Osinski, K. et al., "A 1 μm CMOS process for logic applications," *Philips J. Res.*, Vol. 44, 257-293, 1989.

Pai, P. et al., "Material characteristics of spin-on glasses for interlayer dielectric applications," *J. Electrochem. Soc.*, Vol. 134, No. 11, November 1987.

Pai, P. L. and Ting, C. H., "A framework for multilevel interconnection technology," *IEEE VMIC*, 1988.

Parekh, N. et al., "Plasma planarization utilizing a spin-on-glass sacrificial layer," in L. B. Rothman (ed.), *Proc. Symp. Multilevel Metallization, Interconnection and Contact Technologies,* ECS, Vol. 87-4, 1987.

Parks, H. et al., "Use SRAMs and test structures as yield monitors," *Semiconductor International,* April 1990.

Parks, H., "Yield modeling from SRAM failure analysis," *Proc. IEEE Intl. Workshop on Microelectronic Test Structures,* Vol. 3, March 1990.

Patrick, W. et al., "Application of chemical mechanical polishing to the fabrication of VLSI circuit interconnections," *J. Electrochem. Soc.*, Vol. 138, 1991.

Paul, C.R. and Nasar, S.A., *Introduction to Electrical Engineering*, McGraw-Hill, New York, 1986.

Pecht, M., Beane, D., and Shukla, A., *Singapore and Malaysia's Electronics Industries,* CRC Press, Boca Raton, FL, 1996.

Pecht, M., Dasgupta, A., Evans, J.W., and Evans, J.Y., *Quality Conformance and Qualification of Microelectronic Packages and Interconnects*, Wiley and Sons, New York, 1994.

Pecht, M., Lall, P., and Hakim, E., *Influence of Temperature on Microelectronics and System Reliability*, CRC Press, Boca Raton, FL, 1997.

Pecht, M., Nguyen, L., and Hakim, E., *Plastic Encapsulated Microelectronics*, Wiley and Sons, New York, 1995.

Penning de Vries, R. G. M. and Osinki, K., "Enhanced process window for BPSG flow in a salicide process using a LPCVD nitride cap layer," *ESSDERC*, 1989.

Pennington, S., "An improved interlevel dielectric process for submicron double level metal products," *VMIC*, 1989.

Pliskin, W. A., "Comparison of properties of dielectric films deposited by various methods," *The American Vacuum Society*, 1977.

Pramanik, D., "CVD dielectric films for VLSI," *Semiconductor International*, June 1988.

Radojcic, R. "Reliability Management for Deep Submicron ICs," *Integrated System Design*, September 1996.

Radojcic, R. "Universal Qualification - a Qualification Strategy for Super ASICs," 1992 *IEEE International Wafer Level Reliability Workshop Final Report*, Lake Tahoe, CA, 1992.

Radojcic, R. and Giotta, P., "Semiconductor Device Reliability vs. Process Quality," *Microelectronics and Reliability*, Vol. 32, No. 3, pp. 361-368, 1992.

Radojcic, R., "Generic qualification of semi-custom IC product families," *Microelectronics and Reliability,* vol. 26,no. 3, pp. 471-479, 1986.

Ramakrishna, V. and Harrigan, J., "Defect learning requirements," *Solid State Technology*, January, 1989.

Raupp, G. B. and Cale, T. S., "Step coverage in plasma enhanced deposition of silicon dioxide from TEOS," *VMIC*, June 1989.

Rentelin, P. et al., "Characterization of mechanical planarization processes," *VMIC*, 1990.

Riga, G., "Failure analysis, feedback to integrated circuits design and fabrication," *ATFA 77 IEEE,* September 1977.

Riley, P. E. and Castel, E. D., "Planarization of dielectric layers for multilevel metallization," in L. B.

Riley, P. E. et al., "Limitation of low-temperature low-pressure chemical vapor deposition of SiO2 for the insulation of high-density, multilevel metal very large scale integrated circuits," *The American Vacuum Society*, 1977.

Rothman (ed.), *Proc. Symp. Mültilevel Metallization, Interconnection and Contact Technologies, ECS*, Vol. 87, No. 4, 1987.

Saxena, A. N. and Pramanik, D., "VLSI multilevel metallization," *Solid State Technology*, October 1987.

Saxena, A. N., "Update and future directions for VLSI multilevel interconnection technology," *State-of-the-Art Seminar*, Course Coordinator: T. E. Wade, 1988.

Saxena, A.N. and Pramanik, D., "Planarization techniques for multilevel metallization," *Solid State Technology*, October 1986.

Scalise, J., Honeywell Air Transport Systems Division, private communication, 1996.

Schiltz, A., "Advantages of using spin-on-glass layer in interconnection dielectric planarization," *Microelectronic Engineering*, 5, 1986.

Schmitz, J. E. J., *Chemical Vapor Deposition of Tungsten and Tungsten Silicides*, Noyes Publication, 1992.

Shacham-Diamand, Y. and Nachumovsky, Y., "Process reliability considerations of planarization with spin-on-glass," *J. Electrochem. Soc.*, Vol. 137, No. 1, January 1990.

Singer, P. H., 'Selective deposition nears production," *Semiconductor International*, March 1990.

Singer, P. H., "Depositing high quality dielectrics," *Semiconductor International*, July 1989.

Singh, B. et al., "Deposition of planarized dielectric layers by biased sputter deposition," *J. Vac., Sci. Technol. B.* Vol. 5, No. 2, March/April 1987.

Sivaram, S. and Mu, X. C., "Minimization of dimple size in tungsten filled contacts," *Extended Abstracts*, No. 248, ECS, October 1989.

Sivaram, S. et al., "Overview of planarization by mechanical polishing of interlevel dielectrics," J. Andrews and G. Cellar (eds.), *Proc. 3rd. Intl. Symp. on ULSI Sci. and Tech., ECS*, Vol. 91-11, 1991.

Slay, B., "Best Commercial Practices in Military Semiconductors," *Sharp Proceedings of the 4th Annual Commercial and Plastic Components in Military Applications Workshop*, Nov. 16th, 1995.

Small, M. B. and Pearson, D. J., "On-chip wiring for VLSI: Status and direction," *IBM J. Res. Develop.*, Vol. 34, No. 6, November 1990.

Soden, J. and Hawkins, C., "Electrical properties and detection methods for CMOS IC defects," *IEEE Proc. First European Test Conference*, April 1989.

Stapper, C. et al., "Integrated circuit yield statistics," *Proc. of the IEEE,* Vol. 71, No. 4, April 1983.

Stapper, C. et al., "Evolution and accomplishments of VLSI yield management at IBM," *IBM J. Res. Develop.,* Vol. 26, No. 5, 1982.

Strathman, S. and Lotz, S., "Automated inspection as part of defect reduction program in an ASIC manufacturing environment," *SPIE*, Vol. 1261, *Integrated Circuit Metrology, Inspection and Process Control IV*, 1990.

Streetman, B., *Solid State Electronic Devices*, Prentice Hall, Inc., NJ, 1980.

Su, L., Chung, J., Antoniadis, D., Goodson, K., and Flik, M., "Measurement and Modeling of Self-Heating in SOI NMOSFETs," *IEEE Transactions on Electron Devices,* Vol. 41, pp. 69-75, 1994.

Sum, J. C. et al., "A two micron pitch double-level metallization process, in L. B. Rothman (ed.), *Proc. Symp. Multilevel Metallization, Interconnection and Contact Technologies*, ECS, Vol. 87-4, 1987.

Suzaki, K., *The New Manufacturing Challenge,* The Free Press, 1987.

Sze, S.M. (ed.), *VLSI Technology,* McGraw-Hill, New York, NY, 1986.

Tang, S.M., "New Burn-In Methodology Based on IC Attributes, Family IC Burn-In Data, and Failure Mechanism Analysis," *1996 Proceedings Annual Reliability and Maintainability Symposium*, pp. 185-190, 1996.

Technical note, "Planarization of TEOS oxide interlayers over topography in 150mm wafers," Westech Systems, Inc.

Technical note, "Planarization of thermal oxide sheet films," Westech Systems Inc.

Texas Instruments, "Important Information Concerning User Decision to Procure TI QML Devices in Plastic Packaging Technology," 1995.

Thomas M. et. al. "The mechanical planarization of interlevel dielectrics for multilevel interconnect applications," *VMIC*, 1990.

Thomas, M. J. et al., "A 1,0 μm CMOS two level metal technology incorporating plasma enhanced TEOS," *VMIC*, 1987.

Ting, C. et al., "Spin-on-glass as planarizing dielectric layer for multilevel metallization," in L. B. Rothman (ed.): *Proc. Symp. Multilevel Metallization, Interconnection and Contact Technologies*, ECS, Vol. 87-4, 1987.

Ting, C. H., "Dielectric planarization processes for ULSI," L. Andrews and G. Cellar (eds.), *Proc. 3rd. Intl. Symp. on ULSI Sci. and Tech., ECS*, Vol. 91-11, 1991.

Tompkins, H. G. and Tracy. C., "Tightly bound H2O in spin-on glass," *J. Vac. Sci Technol. B*, Vol. 8, No. 3, May/June 1990.

Tong, J. E. et al., "Process and film characterization of PECVD phosphosilicate films for VLSI application," *Solid State Technology*, January 1984.

Troutman, R.R., IBM Corporation, "Latch-up in CMOS Technology: The Problem and Its Cure," Kluwer Academic Publishers, Boston, MA, 1986.

U.S. Department of Defense Military Handbook 217, *Reliability Prediction of Electronic Equipment,* revision F, notice 2, February, 1995.

Vollmer, B. et al., "Characterization of field oxide transistor instabilities cause by SOG planarization," *Journal de Physique*, Colloque C4, Suppl. No. 9, Vol. 49, September 1988.

Wager, A.J., Thompson, D.L., and Forcier, A.C., Implications of a Model for Optimum Burn-In. *Proceedings 21st Annual Reliability Physics Symposium*, pp. 286-291, 1973.

Wantuck, R., "Fault Coverage: Trade-Off SAF Coverage for IDDQ?" Ford Electronics, *Proceedings of AEC Conference*, 1996.

Warnock, J., "A two-dimensional process model for chemical mechanical polish planarization," *J. Electrochem.* Soc. Vol. 138, No. 8, August 1991.

Watanabe, H. and Todokoro, Y., "Submicron feature patterning using spin-on-glass image reversal," *J. Electrochem. Soc.*, Vol. 137, No. 1, January 1990.

Watson, G.F., " MIL Reliability: A New Approach," *IEEE Spectrum*, vol. 29, pp. 46-49, August, 1992.

Whitwer, F. et al., "Premetal planarization using spin-on dielectric," *VMIC*, June 1989.

Wilson, R. H., "Effect of planarization in VLSI metallization," *Semiconductor International*, April 1968.

Wilson, S. R. et al., "Multilevel interconnects for integrated circuits with submicron design rules," *SPIE Advanced Processing Of Semiconductor Devices II*, Vol. 945, 1988.

Winter, C. and Cook , W., "Interval estimates for yield modeling," *IEEE Journal of Solid State Circuits,* Vol. 4, August 1986.

Wolf, S. and Tauber, R. N., *Silicon Processing for the VLSI Era, Volume 1-Process Technology,* Lattice Press, 1986.

Wolf, S., *"Silicon Processing for the VLSI Era-Volume 2: Process Integration,"* Lattice Press, 1990.

Woo, M. et al., "Characterization of spin-on-glass using Fourier transform infrared spectroscopy," *J. Electrochem. Soc.*, Vol. 137, No. 1, January 1990.

Wu, W. S. et al., "Characterization of SiO2 films deposited by pyrolysis of tetraetylorthosilicate TEOS," *Journal de Physique,* Colloque C4, Supplement to Vol. 49, No. 9, September 1988.

Wu, W. S. et al., "CVD tungsten and tungsten silicide for multilevel metallization," *IEEE Intl. Symp. VLSI Technol. Sys. Appl.*, 1989.

Yamamoto, H. et al., "Reliable tungsten encapsulated AlSi interconnects for submicron multilevel interconnection," *IEDM*, 1987.

Yasuaki, H., 1985, "Oxide Breakdown Reliability Degradation on Si-Gate MOS Structure by Aluminum Diffusion through Polycrystalline Silicon," *Journal of Applied Physics,* Vol. 58, pp. 3536-3540.

Yen, D. and Rao, G., "Process integration with spin-on-glass sandwich as an intermetal dielectric layer for 1.2 μm CMOS DLM process, *VMIC*, 1988.

Yu, M. L. et al., "Surface chemistry of the WF6-based chemical vapor deposition of tungsten," *IBM J. Res. Develop.*, Vol. 34, No. 6, November 1990.

INDEX